NOONOMY

THE TRAJECTORY OF GLOBAL TRANSFORMATION

NOONOMY

THE TRAJECTORY
OF GLOBAL
TRANSFORMATION

**SERGEY
BODRUNOV**

Translated by
William Dubrickson

BOSTON
2023

Library of Congress Control Number: 2023941022

ISBN 9798887193090 (hardback)
ISBN 9798887193106 (adobe pdf)
ISBN 9798887193113 (epub)

Book design by Lapiz Digital Services
Cover design by Ivan Grave

Published by Academic Studies Press
1577 Beacon Street
Brookline, MA 02446, USA
press@academicstudiespress.com
www.academicstudiespress.com

Contents

Foreword   vii

To the Reader: A Door to the Future   viii

Thinkers Who Have Contributed to the Study of the Interconnections
      between Technological and Social Development   x

Step One: Grasping Reality   1

Step Two: Into the World of New Technologies   15

Step Three: To the Threshold of Technological Revolution   31

Step Four: The New Industrial Society's Second Generation   45

Step Five: Civilization at a Crossroads   56

Step Six: Nooindustrial Production   78

Step Seven: Culture as an Economic Imperative   97

Step Eight: From Economy to Noonomy   112

Conclusion: The Path Towards Nootransformation   142

Foreword

———

The theory of noonomy is one of very few attempts to provide a holistic, theoretically grounded perspective on the socioeconomic development of human civilization. For that reason, this theory, whose premises and content have already been articulated by the author in a series of articles and substantial monographs, has attracted attention both in Russia and abroad. But those who (for whatever reason) cannot study these works in full, or who rely on third-hand information, frequently perceive noonomy ideas in a partial, fragmentary, and sometimes distorted way.

This book is conceived as a kind of guide to the fundamental ideas of noonomy theory, making it easier for the reader to understand the theory's underlying approach and come to terms with its logic. The author's ideas are communicated here in the form of concise theses and basic definitions. Step by step, the reader is invited to follow the author's main arguments and conclusions, which are expressed in vivid, concentrated form.

This manner of expression helps to trace the author's series of arguments and conclusions and directly grasp their mutual interconnection. The reader will be able to appreciate how each idea expressed here fits into the overall system of the noonomy worldview and see which points need to be studied further in order to grasp the general theory. This book is based on reworked portions of the author's prior books *Noonomy* (Moscow: Cultural Revolution, 2018) and *General Theory of Noonomy: A Textbook* (Moscow: Cultural Revolution, 2019).

To the Reader:
A Door to the Future

How does one open a door to the future? After all, not all motion signifies movement towards the future: we might be treading water, going in circles, or even turning back, without finding the road that would lead us to a new world, a genuine (and better) future. "If a person does not know to which port they are sailing, no wind is favorable"—this phrase from Seneca is worth remembering every time we think about long-term strategic prospects for development.

To recognize a genuine path towards greater social development, we must go beyond merely using our common sense or orienting ourselves with preconceptions about what would personally benefit us or our neighbors. What is needed is a scientific theory that holds powerful predictive potential; a theory capable of sorting through the facts and tendencies of the present to reveal the possibility, and necessity, of the future.

Developing such a theory requires us to stand upon the stable ground of scientific method. One of the most important components of this methodology is determining the nature of the relationship between technological development and the evolution of the social order. By studying just this fundamental interconnection and mutual influence of human knowledge (as materialized in productive technology) and social relations (which shape the possibility of technological development and, in turn, are affected and changed by technological upheavals) we will be able to make scientifically grounded predictions.

But the future is not a path predetermined by fate. Our actions and decisions will determine whether we wander in darkness, turning away from the prospects we uncover, or manage to take advantage of newly discovered opportunities for development. Theory can tell us which actions are necessary to open the door to the future, but we ourselves must carry out these actions.

The theory of noonomy, developed by the author, offers a view of a path capable of leading us out of the dead ends and contradictions that human civilization has so far managed to generate. Development will always be contradictory: these contradictions are precisely what motivate us to seek solutions to problems and move forwards. It is important that the path leading out of contradiction be creative, not destructive, and that the contradictory aspects of new developmental stages serve as an impulse to improve society rather than plunging it into a morass of conflict.

The author invites you to join him and cast your eyes on a possible path forward, to grapple with the logic of human society's evolution under the influence of greater knowledge and technological progress and appreciate the steps that must be taken in order to move towards a better future.

Thinkers Who Have Contributed to the Study of the Interconnections between Technological and Social Development

---

Aristotle

Posed the question of how technology (*technē*, art or craft) relates to human needs. However, the problem of technology's influence on the ordering of society escaped him, since the slow pace of technological development in antiquity provided no material capable of offering insight into such influence.

Francis Bacon

Following the Aristotelian tradition, Bacon saw the goal of science and technology as providing society with various goods. Formulated the idea that the comparative levels of development of different societies were defined not by the natural conditions of their existence, but instead by their degree of scientific and technological development.

Jean-Antoine de Condorcet

Posed the question of technology's, as well as society's, historical development. Advanced the thesis that scientific progress and the progress of industry influence one another and that their joint impact determines the "progress of the human race."

Charles Fourier

Proposed a theory of stages of social development dependent on the progress achieved by the development of science and production, which he measured according to the possibilities attained for satisfying social needs, on which basis he separated out historical stages with various systems of socioeconomic relations.

Claude-Henri de Saint-Simon

Advanced the thesis that there is a necessary correspondence between economic and political development and a society's intellectual and moral level. Developed a theory of the alternation of critical phases with quiet phases in the development of society and of the interconnection between scientific and political revolutions. He believed that the cause of social development was, in fact, changes in scientific worldviews.

Karl Marx

Theoretically grounded the materialist conception of history and advanced the thesis of production as a twofold process: the production of the goods of life and the production of social relations. Founded a theory of social development as decisively determined by changing modes of production, which appear as a contradictory unity of mutually acting productive forces and production relations, each at a particular level of development.

Vladimir Ivanovich Vernadsky

Articulated humanity's role as the decisive geological force acting on Earth's surface, influencing the development of its biogeochemical shell. Formulated the idea of the noosphere, according to which human reason becomes a factor that regulates the process of natural, as well as social, evolution.

Joseph Schumpeter

Advanced the theory of the entrepreneur as innovator and "creative destruction" (the destruction of old combinations of resources during economic development and the creation of newer, more effective ones). Proposed the idea that Kondratiev cycles depend on the introduction of new technologies and examined the particularities of innovation in different phases of a Kondratiev cycle. Demonstrated the influence of the various phases of a Kondratiev cycle on the course of a medium-term business cycle (a Juglar cycle).

John Kenneth Galbraith

Demonstrated the influence of technological upheavals over the course of the twentieth century on the character of socioeconomic structures (the allocation of property rights, corporate structures, the economy's division into a market system and a planned system, the creation of systems for consciously generating demand, the role of the state, and so on), transformations which he described in his theory of the new industrial state.

Daniel Bell

Proposed the hypothesis of post-industrial society based on the fact that the relative weight of industry in the economies of developed countries substantially decreased during the second half of the twentieth century. Several of Bell's predictions about the evolution of society's economic and social structure have been confirmed. However, many leading economists and sociologists are skeptical of the judgment that contemporary society qualifies as "post-industrial," believing that Bell underrated industry's critically important role and hypostatized knowledge production.

Method of Investigation

Before making our first step along the road of studying the evolution of human society as influenced by technological change, we must define the methodological premises that ground us. The bulk of these premises were worked out during a time when humanity embarked on a path of profound and rapid transformations in the material basis of its existence—the nineteenth and twentieth centuries. These are exactly the transformations that made it possible to discover the laws by which technological change influences social development.

* * *

Karl Marx (1818–1883) was perhaps the first scholar to make a serious contribution toward placing the study of the reciprocal influence of technology, on the one hand, and society's socioeconomic order, on the other, within a firm scientific framework.

* * *

In his research, Karl Marx adopted the point of view that people's social relations are defined by the conditions of their material production and that the interaction of social relations and material production determines the transition from one historical stage of society's development to another. Marx set this principle as the foundation of his theoretical investigations in the realm of political economy, which allowed him to take a step away from formulating general philosophical research principles and towards an understanding of the concrete laws of the mutual influence of technological and economic development.

* * *

The *object of political economy* in Marxism is, primarily, *people's objective relations* within the process of material production in the broad sense of the term, that is, *both* production proper *and* trade, distribution, and consumption. Let me reiterate this: *specifically*, objective social relations, and *specifically*, within production. Marx here continued the tradition of classical political economy. This constitutes the principal **difference** between the science of classical political economy and the current reigning discourse within economic theory (designated by the word "economics") in which attention is centered upon individuals' *subjective* choices and focused primarily on the sphere of *trade* rather than production.

* * *

Political economy always emphasized that it was necessary for development to ensure a *dialectical correspondence between, on one hand, the material and technological basis of production* (which Marxism refers to as "productive forces") *and, on the other hand, socioeconomic relations of production.* Studying capitalism, Marx demonstrated on one hand how the character of economic relations spurs forward technological progress, determining the stages of the evolution of productive forces, and on the other, how this evolution of productive forces modifies that same character of productive relations.

* * *

The significance of the Marxist approach is also defined by its *historical and systematic method of researching economic reality.* For us, the economy is not only a sphere in which various more or less rational actors operate, but a multitude of historically developing economic systems. One of these is Russia's economic system. This system has its own *trends of development* which, in a *specific* (to our region and to our civilization) way, refract the general laws of economic development.

* * *

Without taking Marxism's contributions into account, moreover, we cannot fully understand the *role of economic development's "human dimension"*—for example, the growing importance of **knowledge** for the modern economy. The Marxist method permits us to provide an explanation for the critically important role of science and education in the transition to a model of economic development which would not only make it possible to carry out nationwide *modernization* on a modern technological basis, but also open a path towards *constant* economic modernization based on prioritizing the development of *knowledge-intensive production*—production whose main factor is knowledge rather than machines or raw materials.

* * *

It is not a coincidence that political economy combines two conceptions: politics and economics. Its major strength is that it accounts for the sociopolitical component of economic processes. Indeed, in the framework of Marxist political economy, the most important question is *which* social strata, and in what kind of relationship with one another, are produced by a given economic order.

* * *

In this way, the method of classical political economy not only makes it possible to show the *functional links* between diverse economic phenomena but also it allows us to delineate what is *typical* and what is *contingent* in the economy, to investigate the economy as it *develops*, **systemically and historically**. And, most importantly, by putting processes of material production, rather than processes *of trade alone*, at the heart of its research, it pays very close attention to *human relationships* and social collectives, examining the interaction of *socioeconomic interests*.

* * *

The most valuable thing about Marx for us now is not only the theory of surplus value, nor just the labor theory of value, both of which have been subjected to unending criticism—showing at the very least that these theories, serving as objects of criticism for more than one hundred fifty years, have yet to lose their relevance. Much more interesting for our purposes are *Marx's predictions, which are coming true at this very moment, about material production's long-term developmental tendencies*—predictions that were based, incidentally, on the same theoretical assumptions mentioned above.

* * *

Marx predicted *man's displacement from the direct process of material production*, man's transition from being a direct participant in production into becoming its "watchman and regulator." This is based on the transformation of natural processes into controlled and directed technological processes, "where labor in which a human being does what a thing could do has ceased," when labor appears "as an activity regulating all the forces of nature" and becomes an "experimental science, [a] materially creative and objectifying science," when human development takes place "as a constant suspension of its [own] *barrier*" and amounts to the "absolute working-out of [the human being's] creative potentialities."[1]

* * *

There is another economist known for his research into the problems of technological development and innovation and their influence on society's economic condition and structure—the Austrian-born American economist **Joseph Schumpeter** (1883–1950)—whose ideas undeniably echo Marx's (as Schumpeter himself did not deny). At the same time, Schumpeter substantially differs from Marx in many ways in his approach to studying the role and place of technological upheavals in the economy.

* * *

Above all, Schumpeter introduced the difference between economic *growth* and economic *development* into economic theory. Growth means increased production and consumption of already existing goods, while economic development signifies the continuation of production based on new innovations: new commodities, new technologies, new methods for organizing production.

* * *

Joseph Schumpeter noted that the development of new innovations does not happen continuously over time. He studied the innovative activity of entrepreneurs, leading to technological upgrades in production, as a factor in gaining competitive advantage and as the main engine of economic development.[2]

* * *

He justly viewed competition based on innovation and scientific development by corporations as the main factor underlying the economic dynamics of capitalism.[3] Schumpeter referred to the use of competitive advantage based not on lowering production costs and prices for traditional goods, but rather on new innovations as effective competition.

* * *

---

1 Karl Marx, *Grundrisse*, trans. Martin Nicolaus (London: Penguin, 1993), 705, 325, 612, 712, 542, 488.
2 Joseph A. Schumpeter, *The Theory of Economic Development*, trans. Redvers Opie (New York: Routledge, 2017).
3 Ibid.

Another of Schumpeter's contributions was his differentiation of interconnected complexes of technologies and corresponding stages of technological development. Innovations do not appear gradually and consistently, but come into being as interconnected groups of innovations, that is, as clusters (bundles).[4] It is precisely new scientific discoveries that lead to the emergence of these innovation clusters. The term "waves of innovation" has established itself as a way to refer to periods (stages) of predominant development of particular technologies.[5]

* * *

Joseph Schumpeter tried to combine his theory of the role of technological innovation with N.D. Kondratiev's notion of "long conjunctural waves." In his conception, the phase of a wave of technological change that underlies a Kondratiev cycle interacts with the waves of technological change underlying a medium-term Juglar cycle, with the former determining the latter. In this way, both long-term and medium-term waves of technological renewal and the phases of long-term and medium-term cycles prove to be dependent on one another.[6]

* * *

Schumpeter saw that the concentration of capital in large corporations opens additional possibilities both for financing scientific research and development and for investing in the application of new technologies. At the same time, he feared that the concentration of production and formation of enormous corporate organisms would lead to a decline of individual competitive spirit (along with its romantic urge to innovate), the crowding-out of small and medium-sized entrepreneurs, and the erosion of private property.

* * *

The extent to which Schumpeter's forecasts were and were not validated was later shown by the research of another famous economist, **John Kenneth Galbraith** (1908–2006).

* * *

---

4 S.M. Menshikov and L.A. Klimenko, *Long Waves in the Economy: When Society Changes Its Skin*, 2nd ed. (Moscow: LENAND, 2014), 192.

5 Mark Blaug, *Great Economists before Keynes: An Introduction to the Lives & Works of One Hundred Great Economists of the Past* (Brighton, Sussex: Wheatsheaf Books, 1986), 215–217.

6 Joseph A. Schumpeter, *Business Cycles: A Theoretical, Historical and Statistical Analysis of the Capitalist Process* (New York: McGraw-Hill, 1939), 181–182.

Galbraith observed that economic life was undergoing "the application of increasingly intricate and sophisticated technology to the production of things. Machines have replaced crude manpower. And increasingly, as they are used to instruct other machines, they replace the cruder forms of human intelligence."[7] These processes led to the consolidation of production, demanding more and more significant capital investments, and attracting more and more highly qualified specialists. The result was the emergence of large corporations as the predominant type of economic organization capable of attracting the capital necessary for such production.

* * *

What became quite clear at this point was a process that had begun much earlier: the *fragmentation of the entrepreneur-proprietor figure who both organizes production and reaps its income.* Galbraith, building on the reflections of a series of authors from the first third of the twentieth century (Thorstein Veblen,[8] Adolf Berle and Gardiner Means,[9] Stuart Chase,[10] and others) and in some ways intersecting with Karl Marx's ideas about the separation of joint-stock companies' capital into "capital-property" and "capital-function," noted that at the start of the twentieth century, control by private owners had been replaced with control by the technostructure, that is, managers and technical specialists.

* * *

The growth of corporate capital inevitably led to a transformation in the state's economic role. In the 1960s, Galbraith concluded that "the state undertakes to regulate the total income available for the purchase of goods and services in the economy. It seeks to insure sufficient purchasing power to buy whatever the current labor force can produce."[11]

* * *

The first consequence of these changes was a significant *rise in the role of planning.* "The large commitment of capital and organization well in advance of result requires that there be foresight and also that all feasible steps be taken

---

7   John Kenneth Galbraith, *The New Industrial State* (London: Hamish Hamilton, 1967), 1.
8   Thorstein Veblen, *The Engineers and the Price System* (New York: Viking, 1936 [1921]).
9   Adolf A. Berle and Gardiner C. Means, *The Modern Corporation and Private Property* (New York: Harcourt, Brace & World, 1968 [1932])..
10 Stuart Chase, *A New Deal* (New York: The Macmillan Company, 1932). This book's title was used by Franklin D. Roosevelt to refer to his program in the 1932 election.
11 Galbraith, *The New Industrial State*, 2.

to insure that what is foreseen will transpire"—this was Galbraith's highly important conclusion.[12]

* * *

In addition, *consumer demand became an object of planning.* Galbraith rightly stressed that the nature of technology and its associated capital requirements, along with the time needed to develop and produce products, made state regulation of demand necessary.

* * *

The task of *creating demand* (and not merely keeping track of it) is fulfilled by both the state and—more importantly—corporations. As Galbraith emphasized: "No mechanism of the market relates the decisions to save to the decisions to invest."[13] This assertion clearly echoes John Maynard Keynes' analogous conclusion from the 1930s.

* * *

In the end, Galbraith arrived at the conclusion that there is a profound conceptual distinction between small enterprises, all of whose successes rely on their complete control by a single person, and corporations. This difference, which could be seen as a boundary line distinguishing between millions of small-scale firms and thousands of giants, rests at the heart of the general *separation of the economy into "market" and "planned" systems.*

* * *

Thus, Marx, Schumpeter, and Galbraith all saw that economic evolution is primarily based on technological progress that changes the material foundation of industrial production. From various perspectives, they developed an approach to studying the influence of technological progress on the economic arrangement of society.

---

12 Ibid., 4.
13 Ibid., 42.

Step One

Grasping Reality

1.1. Production and Its Significance

Technology, on the one hand, and people's social relations, on the other, are intertwined very closely in the process of material production. **Material production** is simultaneously the production of the *material conditions* of human social life (human society could not exist if it did not produce) and the production of people's *social relations* and social existence, that is, the production of social people.

* * *

Human beings' social relations within production—the *social arrangement of production*—reflects the state of material production and of people's productive activity. In turn, the social arrangement of production serves as the basis for all other social relationships between people. However, these social relations (social structure, culture, ideology, politics, social psychology, etc.) are not a passive copy of production relations. They, in turn, actively impact the development of the sphere of production.

> *Production* is the process by which humanity transforms what it is given by nature, adapting natural material for human needs, and giving it the required form for consumption.

* * *

Nature can only be transformed once we have understood how it is organized and once the laws of its existence have been discovered. At stake is not only the transformative activity of humanity itself, but the far-reaching consequences of this activity, which also influence humanity's habitat. This is why scientific *knowledge* of the world becomes more significant as the horizon of this knowledge grows longer. Without it, technological improvements would be impossible.

1.2. The Elements of Production (Product, Means of Production, Technology, Labor, Organization, Knowledge)

The process of manufacturing products—that is, the transformation of natural substances for the purpose of satisfying human needs—is known as the production process. The most substantial elements of the production process are human *labor*, human *knowledge*, raw *materials, instruments of labor, technology,* and the *organization* of production.

* * *

The structure of society is affected by every element of the production process: the nature of labor and its level of productivity, the development of knowledge, the means of production that are used (raw materials, equipment, etc.), the technology of production, the character of the product being produced, and—following from all of this—the methods of organizing production.

> The *product* of production is an external object, a thing, which is obtained by transforming natural material via the production process and designed to satisfy human needs.

* * *

A **product** is the materialized result of the application of knowledge (its objectification) towards the satisfaction of human needs—primarily by manufacturing material devices or providing services which rely on the use of material products.

* * *

Products created by human beings have (in the philosophical sense) an objective and thing-like character, but they need not necessarily be embodied as physical objects. Some needs are satisfied by intangible products which do not take substantial form, that is, *services*. There are two factors relevant to services that must always be taken into account.

* * *

Firstly, all services (with very few exceptions) can only be provided with the help of some sort of material product. If material production ceased, there would be no services.

* * *

Secondly, only material products can satisfy humanity's vital needs for food, clothing, housing, transport, communication, processing information, and more. The sole reason that a large contingent of service employees may exist is that other people provide them with necessary material products.

* * *

As production develops, each product created by human beings is less affected by natural substances and more affected by the **technosphere**.

Meanwhile, the technosphere's development is determined not so much by the accumulation of work tools, or how they are used, as by the knowledge embodied in these tools, which determines people's ability to make use of them and amplifies the extent to which humanity may effectively reach its goals: in other words, the *technological application of knowledge and science*.

> The *technosphere* is the totality of technogenically produced objects (made by human beings) and the part of the biosphere that is transformed by these objects. Broadly speaking, the technosphere includes all conditions of productive activity— knowledge, ability, people's mutual relations within the process of production, etc.

* * *

Technological development determines the evolution of production, as well as its very product. To define the extent of this evolution, we must introduce the concept of the **level (of complexity) of a product.**[1]

---

1 For more on this concept, see S.D. Bodrunov, *The Future. The New Industrial Society: A Reboot* (Moscow: Cultural Revolution, 2016), 13–14.

The *level (of complexity) of a product* refers to the characteristics of a produced product that determine its multiple stages of processing and the corresponding volume of knowledge that goes into it.

* * *

This concept may be expressed in purely quantitative terms, defining how many stages of processing have been undergone by raw materials to convert them into a final product that satisfies particular needs. Much more important, however, is the qualitative appraisal of a product's complexity.

* * *

Philosophically speaking, any given product is constituted by objectified human knowledge, which is applied to produce a thing or object that embodies it. The general tendency of productive development is to lower the need for natural substances, natural energy, and natural forces in manufacturing a product. This leads to decreased specific consumption of raw materials. At the same time, progressively more complex equipment comes to play a much greater role in structuring products, along with (most importantly of all) the increasing extent of knowledge needed to deliver a higher-level product.

* * *

The active force that unites all the component parts of production in a single process is human **labor**.

Labor is practical human activity oriented towards changing natural objects for the satisfaction of human needs.

* * *

In the words of Karl Marx,

> . . . Labor is, first of all, a process between man and nature, a process by which man, through his own actions, mediates, regulates, and controls the metabolism between himself and nature. He confronts the materials of nature as a force of nature. He sets in motion the natural forces which belong to his own body, his arms, legs, head and hands, in order to appropriate the materials of nature in a form adapted to his own needs. Through this movement he acts upon external nature and changes it,

and in this way he simultaneously changes his own nature. He develops the potentialities slumbering within nature and subjects the play of its forces to his own sovereign power.[2]

* * *

Labor, then, is practical activity: a person engaged in labor pursues a definite goal and directs their efforts towards obtaining a definite final result. To accomplish this, the laborer should clearly know what he wants to achieve, that is, mentally formulate an image of the final product. Moreover, he must imagine how, and using what *technologies*, the desired result can be obtained. This requires specific knowledge. Besides knowledge, a laborer must have the skills and abilities needed to put his mental image into practice. Within the labor process, a not insignificant role is played by the laborer's capacity to bend his will toward the goal he is striving for; to concentrate and mobilize his knowledge, abilities, skills, and energy to deliver the final result—the product of labor.

* * *

In its simplest definition, **technology** designates the totality of productive methods and processes that are used to process raw materials into a finished product.

> *Technology* is the mode of interaction of all the elements of a concrete production process, oriented towards making a finished product. It is facilitated by the set of abilities and knowledge required to make this product.

* * *

Without technological knowledge, without the knowledge required to develop and use technologies, no normal production process capable of reaching a posited goal can be ensured.

* * *

Analyzing the fundamental components of the production process in the modern industrial mode of production and studying the correspondence between the service sector and material production leads to the conclusion that technological development (or the technological application of scientific

---

2 Karl Marx, *Capital: A Critique of Political Economy*, vol. 1, trans. Ben Fowkes (New York: Penguin Books, 1982), 283.

knowledge) plays the decisive role in forming technological paradigms that unite all the components of the production process at a given stage of development. Technology permeates the entire production process—from the choice (or creation) of original materials, to the quality of living labor and the means of production that are used, to the organization of the production process and completion of the finished product.

* * *

To a significant extent, technology determines how the industrial mode of production affects other levels of the social structure, changing the face of society itself. Of course, this influence is complex, mediated, and multilayered, interacting with the laws of development of various social spheres and strata. Yet it is the impulse originating from technological progress that plays the decisive role.

* * *

The **organization of production** takes on exceptional importance in the industrial mode. This is due to two important factors. Firstly, understanding the growing complexity of production, the interaction of its many components, and the need for rational management requires deep and specialized knowledge. The second factor is the conversion of production from a primarily individual into a primarily collective process, based on the combined labor of many people. Managing people's combined activity within industrial production through specialization and division of labor is a crucial aspect of organizing production.

* * *

The development of methods for organizing production was somewhat belated compared to the other elements of the industrial mode of production. True, methods of linear organization were emerging more or less spontaneously even in the nineteenth century, based on spatially distributing machines and mechanisms along linear tracks that allowed for step-by-step operations to be performed on basic materials and component parts. But we observe the first consciously developed methods for organizing production only at the start of the twentieth century: Taylorism, Fordism, and the assembly-line method of manufacturing.

* * *

As technology, labor, and products grow ever more complex, organization plays a progressively greater role in increasing productive efficiency. Each step taken to improve industrial technologies demands a corresponding improvement in productive organization, aimed at the more and more efficient provision of industrial outputs (both goods and services). Meanwhile, the complexity of productive organization depends significantly on how much knowledge is involved in the development of the methods of organization used to make the production process function.

1.3. Production's Stages of Development

In order to come to terms with production's various stages of development, a historical perspective on the problem is needed above all. Only a conception of the historical development of both the technosphere and human society, in their interaction and interdependence, will permit the objectively necessary laws determining how transitions occur from one developmental stage to another to emerge.

* * *

The method of classical political economy not only makes it possible to demonstrate the functional links between disparate economic phenomena but also allows us to distinguish between what is regular and what happens by chance in the economy, and to examine the economy in its systematic and historical development. Most importantly, by analytically centering the processes of material production, and not merely those of trade, it pays rigorous attention to the relations between people and social groups, examining the interaction of different socioeconomic interests.

* * *

Different socioeconomic arrangements of society correspond to different stages of the development of the production process. The transition from an appropriative economy, reliant mainly on stone tools, to a production-based economy using metal tools led to the development of the division of labor (the division between arable farming and animal husbandry, the separation of trades from agriculture, etc.) and served as the key pivot point away from a purely natural economy to a commodity economy. Meanwhile, the production of surplus product, appearing at the same time, facilitated the development of various forms of exploitation.

Social division of labor is the specialization of producers on a society-wide scale, based on specific types of activity. It is a prerequisite for the phenomenon of exchange of the products of specialized production between producers to emerge.

A *natural economy* is an economic form in which goods produced within a closed, localized economic community (a family, a commune, an agricultural estate) are also consumed within that community. Some part of productive output may be redistributed beyond the community's boundaries, but not through trade.

A *commodity economy* is an economic form based on the fully developed division of labor: each producer, specializing in the independent, private production of specific goods, receives the means for sustaining their own life (and their productive labor) by trading their products for others on the market.

Surplus product is that part of productive output which exceeds what is necessary for the producer's own direct consumption. The production of surplus product assumes that a certain level of labor productivity has been achieved.

Exploitation is the coercion of one human being by another (the exploiter) into labor, in order to produce surplus product which is appropriated by the exploiter.

* * *

The transition from primarily manual instruments of labor to machines occurred due to the specialization of tools and the use of universal heat engines that did not depend on natural energy sources such as water, wind, or the muscular strength of humans and animals. This laid the foundations for the transition to a new mode of production. As machine technology spread far and wide, commodity exchange, the circulation of money, and the development of capitalist social relations were energized and given a new basis for far-reaching expansion, leading to the hegemony of the *capitalist mode of production* and the development of the *world market*.

* * *

Machine production brought enormous progress in the division of labor, a hitherto unknown rise in productivity, the development of a variety of new needs, and the discovery of all sorts of new ways to satisfy them. All this progress encouraged the illusion of "mastery over nature." Humanity's supposed mastery over nature is a one-sided narrative: humanity cannot interact with nature without considering the principles according to which the natural environment reproduces itself. Acting against these principles does not bring any reward, but rather, harm, and sometimes even catastrophe.

* * *

The rise of machine production was the first stage in the development of industrial production.

> *Industrial production* is the production of standardized instruments for satisfying human needs on a massive scale, based on the use of intricate tools that give humanity control over natural processes (mechanical, physical, chemical, biological) which are transformed by technological processes, as well as the use of artificial sources of energy.

* * *

Since it is able to satisfy human needs on a mass scale, industrial production also commands the technical capacity to satisfy individualized needs. Modern industrial production relies on both machine and non-machine technologies, built for human control over various non-mechanical processes—physical, chemical, biological, and informational. Industry serves as the robust technological core of the modern economy. It is the evolution of industry that has largely determined the shifts in the socioeconomic order of society over the last two hundred fifty years.

* * *

The scholars who have most contributed to the study of the laws of interaction between humankind's technological and socioeconomic development have done so precisely by investigating the industrial mode of production. The nature of these scholars' direct contributions to economic theory is the subject of heated debate, but what matters for our purposes is the philosophical aspect of their approach to studying socioeconomic reality, and above all, the research methods of classical political economy.

1.4. Industrialization and the Industrial Revolution

The transition to the industrial mode of production—the **Industrial Revolution**—was marked by deep transformations in the technological basis of material production, along with equally substantial changes throughout the entire system of social relations.

> The *Industrial Revolution* was the transformation from artisanal manufacturing—based on manual labor, the use of hand tools, and natural energy sources—to industrial production.

* * *

Over the course of the Industrial Revolution, the material basis of production changed due to the following qualitative shifts:

- the transition from primarily hand-held tools to complex equipment that combines an energy source, labor process (an instrument of labor)—that is, a means of mechanical, physical, chemical or biological action upon an object of labor, and transformative control devices that give the labor process a necessary and expedient form;
- the transition from relying on the muscular strength of humans or animals, along with natural sources of energy (water and wind), to universal artificial energy sources.

* * *

The first stage of the Industrial Revolution took place above all through the development of large-scale machine production.

> A *machine* is a composite piece of equipment based on the use of mechanical processes, consisting of an engine (powered by heat or electricity), labor tools that act upon the object of labor, and a transfer mechanism that communicates a necessary practical form to the action of the labor tools.

The transition from production based on manual labor and hand-held tools to industrial production had the following technological consequences:

1. abrupt growth in labor productivity;
2. the ability to standardize production on a mass scale;
3. independence of production sites from natural energy sources;
4. independence of production sites from local sources of raw materials, due to the development of machine transportation;

5. reliance on scientific research for converting natural processes into technological ones, which opens the possibility of constant technological advancement.

* * *

The Industrial Revolution enabled the complete **industrialization** of society's economic system.

> *Industrialization* of the economy is the expansion on a mass scale of standardized production, originally developed within the industrial sector, to all other areas of the economy (transportation, communications and other branches of the service sector, construction, agriculture). During industrialization, industry's economic importance grows, and industry becomes the economy's technological core.

1.5. Post-Industrial Society?

> *Post-industrial society* is a theory stating that the growth in the service sector's relative share of GDP in developed countries, and the reduction of industry's share, signal society's transition to a post-industrial stage of development in which material production loses its significance compared to the production of services, particularly knowledge and information.
>
> "The 'post-industrial' society . . . [is] one in which the economy [has] moved from being predominantly engaged in the production of goods to being preoccupied with services, research, education and amenities . . ."[3]
>
> "The change in the meaning of knowledge that began 250 years ago has transformed society and economy. Formal knowledge is seen as both the key personal resource and the key economic resource. *Knowledge is the only meaningful resource today.* The traditional 'factors of production'—land (that is, natural resources), labor and capital—have not disappeared. But they have become secondary."[4]

* * *

---

3   Daniel Bell, "Notes on the Post-Industrial Society (II)," *The Public Interest*, 7 (Spring 1967): 102.
4   Peter F. Drucker, *Post-Capitalist Society* (New York: Routledge, 2011), 38.

We cannot say that these accounts of post-industrial society were completely detached from reality. Certain developmental tendencies were observed quite correctly by the post-industrialists, but they often drew inaccurate conclusions from these tendencies. For example, they conflated the *relative decline* of material production and the *relative decline* of material expenditures in the cost structure of producing products with a *decline in the role of* material production. They were inclined to project the genuinely profound changes in the lives of a relatively narrow stratum of workers within the modern information, telecommunications, and media sectors onto the greater part of society.

* * *

The post-industrialists fetishize knowledge and information, and the role of the creators, transformers, and distributors of such knowledge and information, by treating them as intrinsically significant and isolating them from the real process of material production. A myth of the "new economy" has emerged, which supposedly offers developed countries a path towards crisis-free growth. Post-industrial theory factually ignores the significance of the industrial basis of society's existence and development. In our opinion, this is post-industrialism's primary mistake, since so-called "knowledge production"—and the production of information, media products, and means of telecommunication—is only truly significant insofar as it aids the functioning and development of material production.

* * *

Not only are theories of post-industrial society premature in their assumption that tendencies affecting only a small segment of the economy pervade nearly the whole of society's production. More importantly, they ignore the role of such fundamental factors in social development as material production and its central core, industry, along with immaterial factors such as human culture; meanwhile, it is these exact factors that predetermine the level and character of human activity's development, which means that they determine humanity's social existence as a whole.

* * *

Marx, who predicted the colossal growth of the role of knowledge in production even from the vantage point of the mid-nineteenth century, was much more prescient than were the late twentieth century's theorists of post-industrialism. Without singing the praises of science and knowledge "as such," he clearly indicated the significance of science and knowledge precisely in their technological and productive applications. *Society needs knowledge,* not for "finding creative solutions," but *for the development (through the application of knowledge) of modern material production that satisfies* the genuine *needs* of humankind.

* * *

Note that for all the changes in material production over the last hundred years, it is industrial technology that remains the foundation of our economic existence, which rests above all on industrial production. The latter is what enables the uninterrupted growth of labor productivity within material production, based on scientific and technological progress, and creates possibilities for the growth of employment within the service sector.

1.6. The Role of Knowledge

Since their first appearance, industrial technologies have required ever-greater applications of scientific *knowledge.*

* * *

It was necessary to study the properties of various materials in order to develop practical methods for processing them and create materials with the desired characteristics, and to study the properties of various types of energy (mechanical, thermal, electrical) in order to understand how to use them, convert them, transmit them, and use them in the process of production. Extensive research was required to make possible the creation and use of complicated machines, and to interfere in the complex physical and chemical processes that take place when original materials are processed using various tools. Finally, the labor process itself became an object of scientific research, in pursuit of more effective ways to make use of humanity's capacity for work.

* * *

One might say that the level of a technology directly depends on the extent of the knowledge concretely implemented in it. The common term "high-tech" refers precisely to the technologies that make use of cutting-edge scientific knowledge.

* * *

The complexity and endless diversity of industrial technologies have developed the specialization and division of labor to a tremendous degree. Consequently, most participants in the production process no longer have full command over the technologies used to produce any given product. Instead, they perform only one specific and partial function within the framework of a particular technology.

* * *

In the final analysis, it is precisely the character and the extent of the *knowledge* embedded in a product that determines its *level*. Any consumer features we might impart to a product, and all the technical characteristics of its use, depend on the knowledge used to create it. The augmentation of knowledge used in production guarantees the augmentation of products' capacity to respond to human beings' ever more diverse demands.

* * *

The greater or lesser potency of the knowledge contained in a product correspondingly leads to the greater level (increased complexity) or lesser level (decreased complexity) of that product. In the exact same way, growth in the knowledge-density of technologies leads to their advancement, while a decrease in knowledge-density leads to their becoming more primitive; the augmentation of a worker's knowledge leads to a growth in his or her qualifications, while a lower level of knowledge leads to professional deskilling.

* * *

Because of the tremendous role of knowledge in the industrial mode of production, since the nineteenth century there has been an observable tendency for production to transform itself and toward the transmission of knowledge, that is, its technological application in specialized branches of social production. Developments in science, education, and professional training take up an ever-greater share of state budget expenditures and gross domestic product. The sphere of knowledge production and transmission interacts more and more closely with the sphere of direct production.

Step Two

Into the World of
New Technologies

2.1. Technological Paradigms

Technological development within the framework of economic industrialization leads to substantial qualitative changes not only within production, but in society's entire way of life. A certain amount of accumulated change over a particular time period brings about a transformation that presents society with a new level at which needs can be satisfied. We must find criteria for distinguishing between qualitatively different periods of technological development that determine qualitative distinctions in the level of society's needs, as well as the means and capacity for satisfying them.

*　*　*

The question of how to identify interconnected technological networks and their corresponding levels of technological development carries with it a long scholarly history. Joseph Schumpeter already observed that the development of new innovations occurs discontinuously. He identified networks of new technologies that fostered surges of innovation, which he dubbed "technology clusters."[1] The periods in which one technology cluster was replaced by another were referred to as "waves of innovation."[2]

* * *

In 1975, the West German scholar Gerhard Mensch researched how periods of technological stagnation, characterized by the preponderance of steadily improving or even purely imaginary innovations, are replaced by periods in which fundamentally new technological solutions become entrenched.[3] In the 1970s and 1980s, the English economist Christopher Freeman formulated the concepts "technological system" and "techno-economic paradigm," which were further developed by his student Carlota Perez.[4]

* * *

The term *technological paradigm [tekhnologicheskiy uklad]*, as used in Russian economic scholarship, is an analogue of concepts such as waves of innovation, techno-economic paradigms, and technical modes of production. It was first proposed in 1986 by Dmitriy Lvov and Sergey Glazyev.[5]

A *technological paradigm* is a system of interconnected forms of production (including interdependent technology chains) with a common technical level. These may also be considered as subsystems of a broader system of social production that includes multiple technological paradigms.

* * *

According to Sergey Glazyev's approach, a technological paradigm is a cohesive and stable framework within which a self-contained cycle takes place, beginning with the obtainment of initial resources and concluding with the output of final products that correspond to a type of social consumption. The core of a technological paradigm is a set of basic technologies that are used throughout a fairly long period of time. The technological innovations that define how this core is formed are called key factors. Industries that intensively

---

1 Menshikov and Klimenko, *Long Waves in the Economy*, 192.
2 Blaug, *Great Economists before Keynes*, 215–217.
3 Gerhard Mensch, *Das technologische Patt: Innovationen überwinden die Depression* (Frankfurt am Main: Umschau Verlag Breidenstein, 1975).
4 See Carlota Perez, *Technological Revolutions and Financial Capital: The Dynamics of Bubbles and Golden Ages* (Northhampton, MA: E. Elgar Pub., 2002).
5 See D.S. Lvov and S.Y. Glazyev, "Theoretical and Application Aspects of the STP Management," *Economics and Mathematical Methods* (1986).

use these key factors and play a leading role in the dissemination of the new technological paradigm are known as carriers [*iavliaiutsia nesushchimi*].[6]

* * *

Moments of change over the course of historical development are due to the movement from one technological paradigm to another, which can be connected to Nikolai Kondratiev's "long waves of the economic conjuncture." We may note that while the appearance and dissemination of a new technological paradigm coincides with the upward phase of a Kondratiev cycle, the paradigm lives on even after the wave that generated it has passed by and been replaced by another. Modern economic literature usually refers to six technological paradigms.

The *first technological paradigm* (1770–1830) was formed by the expansion of machine technologies within the textile sector. Textiles were its primary industry.

The *second technological paradigm* (1830–1880) was linked to the use of steam engines and the development of railroad transport and transcontinental steamship communication. Many areas of production were mechanized. Its primary industries were the production of railway equipment, steam engines, and steel.

The *third technological paradigm* (1880–1930) was characterized by the development of electrical energy and the internal combustion engine; the development of heavy machinery, electrical engineering, and the growth of the aviation and automobile industries; and the use of the radio, telephone, and telegraph as means of communication.

The *fourth technological paradigm* (1930–1980) was based on the widespread application of the internal combustion engine, powered by oil, petroleum products, and gas; the development of petrochemical technologies; and the appearance and broad use of synthetic materials. Computers and their software appeared; space exploration began.

The *fifth technological paradigm* (lasting from the beginning of the 1980s until the present) is characterized by the wide dispersal of information and communications technologies,

---

6 S.Y. Glazyev and V.V. Kharitonov, eds., *Nanotechnologies as a Key to a New Technological Structure of the Economy* (Moscow: Trovant, 2009), 11.

based on developments within the fields of microelectronics and computer science. Newly developing sectors include biotechnology and genetic engineering, robotics, fiber-optic systems, and space telecommunications.

In the first decade of the twenty-first century, the transition to the *sixth technological paradigm* began, which (it is assumed) will be characterized by the widespread expansion of biotechnology, nanotechnology, and various non-machine or hybrid technologies, such as cognitive technology.

* * *

Various scientific sources allow us to distinguish key technologies and industries that constitute the core, and chronological frame, of different paradigms. The schema of technological paradigms offered here is not a dogmatic or essentialized one. What matters is that each of these paradigms is a coherent technological system, and that the core of each paradigm unites all its components as links in a technological chain. These links' degree of technological and economic cohesion determines the functional effectiveness of the technological paradigm, and the speed at which new technologies are transferred between industries and regions.

* * *

Each paradigm serves as the foundation of a new stage of social development, becoming the key factor for change. However, we must not fall into the trap of technological determinism by extrapolating new stages of social development *directly* from technological transitions and merely reducing all social transformations to the influence of technological innovation. Social orders, even when only their socioeconomic systems are considered, are far more complex; we can only understand the changes occurring within society by drawing from well-developed scholarly methods. For our purposes, the most significant of these comes from political economy.

2.2. Technological Revolutions and Industrial Revolutions

The transition from one technological paradigm to another may be considered a technological revolution. And indeed, such a transition signals substantial qualitative shifts in the technological basis of production: the replacement of certain fundamental technologies within the most important industrial processes by others which are more modern and effective.

* * *

Since technology encompasses all the components of the production process, the shifts mentioned above occur within the content of work, the structure of employment, the systemic use of the means of production, the knowledge required for production, and the organization of production. For some time now, the most important role in technological development has been played not just by knowledge, but specifically by scientific knowledge. Consequently, "knowledge production" has become its own separate sector of the economy. For this reason, beginning with the fifth technological paradigm, the transition to a new paradigm and the formulation of its basic assumptions have been viewed as nothing less than a scientific and technical revolution.

* * *

Technological revolutions are based not only on the creation and distribution of new technologies, nor just the expansion of these technologies' scope to cover all the component elements of the production process, but also the technological transfiguration of the core of modern production—industry. Not all technological revolutions may be called industrial revolutions, but only those which qualitatively transform the foundational technologies of industrial production.

* * *

The first industrial revolution was the general transition from pre-industrial to industrial production. Later industrial revolutions transformed the face of already existing industries. The second industrial revolution was marked by movement towards industrially producing not just objects of consumption, but also the means of production themselves. By now, we are already experiencing the maturation of the fourth industrial revolution, based on technologies that—for the first time—allow production to occur with practically no human involvement.

* * *

One of the preconditions of the fourth industrial revolution was general consciousness of the modern economy's industrial core. Enthusiasm for post-industrial theories corresponds to the trend of developed countries, chasing the potentialities opened by globalization, transferring their industrial production to countries with low labor costs. Within their territory, developed countries have preserved only those links of global "value chains" that foster the creation of knowledge: research and development, organization and management of logistics flows, marketing and advertising. This has led to developed countries' *deindustrialization*.

> *Deindustrialization* (of a national economy) is the reduction of the relative economic weight of industry, to the extent that this is not just caused by the faster development of other economic sectors. Deindustrialization is the consequence of either an absolute reduction in the output of industrial products, or the offshoring of this output to foreign countries.

* * *

The crises that have confronted developed nations since the onset of the Great Recession of 2008–2009, and the significantly strengthened position of many developing countries in high-tech economic sectors, have shown that offshoring is fraught with the danger of making developed countries dependent on other nations for their industrial goods, including high-tech products. Developing countries have gradually begun to move away from the role of subcontractors or "assembly plants" for transnational corporations and towards the creation of their own high-tech productive capacity, based on "reverse engineering" and the accelerated development of their R&D sectors.

* * *

It is becoming clear that merely maintaining control over "knowledge production" is insufficient for sustainable development, and that the "knowledge economy" itself may only develop fully in tandem with the high-tech industrial core of the national economic system.

Because of this, and in order to overcome the negative consequences of deindustrialization, the world's developed nations are beginning to develop and implement policies of *reindustrialization and reshoring*.

> *Reindustrialization* is the restoration of priority to industrial and material production within the economy, coming after processes of deindustrialization.
>
> *Reshoring* is when productive capacity that had been relocated abroad is returned to its country of origin, or the creation of new capacity to replace what was lost.

2.3. The Sixth Technological Paradigm

Nowadays we are moving towards the sixth technological paradigm: the world of biotechnology, nanotechnology, robotics, new medicine that will significantly increase people's life expectancy and quality of life, virtual reality technology, etc. The contours of the technologies that are destined to form the basis of the economy of the future are coming into view. Experts estimate that if the current tempo of techno-economic development is maintained, the sixth technological paradigm will crystallize around the year 2025 and reach its maturity in the 2040s. Meanwhile, from 2020–2025, the next *scientific, technical, and technological revolution* will enter its decisive phase. The basis of this revolution will be new discoveries that synthesize the accomplishments of the fields mentioned above (and, perhaps, others too).

> The *sixth technological paradigm* is a technological complex that is currently forming, which includes nano, bio, information, and cognitive technologies. Its distinctive feature is technological convergence and the formation of hybrid technologies integrated by IT ("digitalization," artificial intelligence, data processing on a massive scale).

* * *

When determining strategies of industrial development, one must remember that transformations in material production will be *systemic, cohesive,* and *interconnected.* Let us highlight some of the *key* changes that must be considered when *creating a new industrial system that matches the cutting edge of twenty-first-century science and technology.*

* * *

The main particularities of industrial development in the near future will include:

- updates in the content of technological processes;
- changes in the (micro-level) structure of industrial enterprises;
- changes in the (macro-level) sectoral structure of industry;
- different approaches to the organization/localization of production facilities;
- the formation of new types of industrial corporations;
- strengthened integration of production with science and education;
- the transition to an ideology of "continuous" innovation in production;

- the formation of economic relations and institutes oriented towards industrial and techno-scientific progress.

* * *

The following must all be renewed: the *content of technological processes;* the *structure* of industries and *location* of production facilities; the internal *structure* and *types of cooperation of production facilities* and their integration with science and education; economic *relations* and *institutions* that ensure the progress of material production according to new principles.

* * *

We must not limit ourselves to the mastery of technologies that manufacture the products we currently need. We must propagate *new standards* in product quality control, industrial management, logistics, and personnel work. *These changes should affect every part of the production process:* its *organization, technological base, output,* and, of course, the *character* and *quality* of industrial labor. For example, so far as transformations in the *character* and *organizational form* of industrial production are concerned, it would be worth attending to the tendency towards the *individualization of production,* which has been apparent since the end of the twentieth century, and the *consumer-to-business* model of labor organization.

* * *

The main technological challenges for industry in the twenty-first century include:

- the accelerated creation of new technologies that raise productivity and lower production costs;
- furthering the "individualization" of the production process, applied technology, and manufactured products;
- implementing the principle of modularity in production;
- speeding up the intellectualization, computerization, and robotization of production;
- developing networked technologies and implementing the network principle of productive organization;
- the miniaturization/compactization of production;
- strengthening the tendency for production sites to be cost-effective and waste-free;

- permanently raising the tempo of technology transfer;
- increasing the "physical" proximity of designers and manufacturers to speed up the implementation of new product ideas;
- the expansion of "zones of intellectualization" in the labor process;
- the "clustering" of industrial relations;
- giving a greater role to the individual, motivational, psychosocial, and other characteristics of the participants in productive activity;
- lowering labor costs for the production of new products while the costs of researching and developing them increase;
- transforming the yield structure of production in favor of knowledge-intensive and high value-added products.

* * *

Most significant is the principle of the simultaneous individualization and *modularization* of production for high-tech spheres such as machine-tool manufacturing, aircraft construction (both civil and military), heavy engineering, and more. Individualizing production and establishing contact between the producer and the individual consumer is in line with the use of modern telecommunications and information technology. The development of the internet led to the formation of massive platforms fostering B2B and B2C communication. Consequently, an effective toolkit was created for direct interaction between clients (consumers) and producers. When combined with the extensive development of technologies based on new principles (virtual engineering, computer visualization, 3-D printing, etc.), this will soon make it possible to *individually* create industrial products that are practically *waste-free* and can be almost *instantaneously delivered* to consumers.

* * *

At the same time, the individualization of production facilitates the transition to *networked principles of organization, not just for businesses, but for the entire process of material production.* This makes it possible to operatively create and change the configuration of producers' interactions with subsidiaries such as subcontractors and outsourcers. In this way, it is possible to swiftly adjust manufactured products based on consumers' individual requests, and then move on to new products oriented towards other consumers, other users, other markets, and so on. In turn, networked organization facilitates the ever more expansive individualization of production; as these processes reinforce each other, they pick up speed like an avalanche.

2.4. Technological NBICS-Convergence and Hybrid Technology

A unique feature of the sixth technological paradigm is not only the increased knowledge-intensity of products, but also the interaction of multiple kinds of knowledge and, therefore, multiple kinds of technology in the production of any given product. Most important in this regard is the integration/convergence/ interconnection of information technology, biotechnology, nanotechnology, and cognitive science. This phenomenon has been called *NBIC-convergence* (N for nano, B for bio, I for info, C for cogno). The term was introduced in 2002 by Mihail Roco and William Bainbridge, who produced the most significant work on this topic: namely, the report *Converging Technologies for Improving Human Performance*, prepared for the World Technology Evaluation Center (WTEC).[7]

> *NBIC-convergence* is the mutual interpenetration of nanotechnologies, biotechnologies, information technologies, and cognitive technologies, leading to the creation of technological processes in which these technologies functionally condition one another and form a contiguous whole.

* * *

The report also proposed the concept of NBICS-convergence, which adds the social sciences to the mix.[8] Though this approach has found widespread use in both Western and Russian scholarship, the social sciences have yet to make a noticeable substantive contribution to solving the problems of developing and applying convergent technologies.[9] As of yet, sociotechnology has only found a real application in the development of artificial intelligence systems meant to interact with their user (or rather, to manipulate their user). Humanistic disciplines spend more time writing about social problems in light of new

---

7   This report defines NBIC-convergence as the "synergistic combination of four major 'NBIC' (nano-bio-info-cogno) provinces of science and technology." See Mihail Roco and William Bainbridge, eds., *Converging Technologies for Improving Human Performance: Nanotechnology, Biotechnology, Information Technology and Cognitive Science* (Dordrecht: Springer, 2003), 1.

8   See J. Spohrer, "NBICS (Nano-Bio-Info-Cogno-Socio) Convergence to Improve Human Performance: Opportunities and Challenges," in *Converging Technologies*, ed. Roco and Bainbridge, 101–117.

9   See M.V. Kovalchuk, Convergence of Science and Technology—a Breakthrough in the Future, *Nanotechnologies in Russia*, vol. 6, No. 1–2 (2011), 21; and M.V. Kovalchuk, O.S. Naraikin and E.B. Yatsishina, "Convergence of Science and Technology and Creation of New Noosphere," *Nanotechnologies in Russia*, vol. 6, No. 9–10 (2011), 10–13.

technologies than proposing ways to integrate social knowledge in technological development.

* * *

Taking into account the sixth paradigm's interconnected technologies, as well as the interdisciplinary character of present-day science, we may (in the long term) expect the merging of the NBIC fields into a single scientific and technological realm of knowledge.

This realm's object of study and action will be the organization of matter at nearly every level: from molecular substances (nano) to living ones (bio), reason and rationality (cogno), and the processes of information exchange (info).

* * *

Thus, the distinguishing features of NBIC-convergence are as follows:

- the intensive interaction of different areas of science and technology, with the effect of significant synergy;
- the broad scope of these mutually influential fields, from matter at its atomic level to intelligent systems;
- emerging horizons for the qualitative growth of technological capacities for humanity's individual and societal development.[10]

* * *

Technological convergence within the framework of the sixth paradigm has led to the broad dissemination of hybrid technologies, in which various combinations of machine and non-machine technologies are used (together with information technologies) as an instrument for regulating and directing natural processes to reach desired goals, opening the door to a new technological revolution.

Hybrid technology refers to the combination of two or more types of technology within a single system to achieve a particular useful result. Such combinations are often **convergent**, meaning that one technology somehow enables the functioning of another.

10 Valeria Pride and D.A. Medvedev, "Phenomenon of NBIC-Convergency: Reality and Expectations. Russian Transhumanist Movement," accessed October 15, 2022, http://transhumanism-russia.ru/content/view/498/110/.

Searching Google for "hybrid technology" on March 31, 2019, returned 641,000,000 results (the Russian term *gibridnye tekhnologii* returned 17,900,000 results). These mentioned hybrid technology in the context of industrial processing, auto manufacturing, pre-sowing seed treatment, electronic system security, atomic desalination plants, military affairs, machine translation, heart surgery, and more. It is hard to imagine an area where hybrid technologies could not be used. However, there was no general definition of the concept to be found in the Russian-language results. The English-language results included one website, dedicated to climate technology, that offered a definition: "Hybrid technology systems combine two or more technologies with the aim to achieve efficient systems."[11]

2.5. Additive and Distractive Technologies

Despite significant growth in the role and importance of non-machine technologies (bioengineering, etc.) the sixth technological paradigm still remains within the framework of the industrial mode of production. Convergent (hybrid) technologies, in fact, give the industrial mode of production a "second life," by uniting machine-based and non-machine-based principles of action upon nature to create products that satisfy human needs with minimal consumption of materials.

* * *

Significant possibilities are opened by technology (for example, 3D printing) that integrates modern machine techniques (printing) with information technology that virtualizes and digitizes reality (3D modeling). This may lead to sharp growth in the expansion of additive technologies and a relative decrease in traditional industrial manufacturing. Instead of "processing" initial materials using distractive ("subtractive") productive technologies (by cutting, grinding, or sawing material from workpieces), newer processes "assemble" products from their component parts by combining or building up materials (usually layer by layer) to create an object from a 3D model.

* * *

11 "Hybrid Technology," CTCN, UN Climate Technology Centre & Network, accessed October 15, 2022, https://www.ctc-n.org/technologies/hybrid-technology.

There are some traditional industrial technologies that might also be considered additive: for example, casting or sintering construction materials, or powder metallurgy. Today, the possibilities afforded by these technologies can be combined with those afforded by 3D printing. We are already witnessing the creation of 3D printers capable of printing entire buildings and structures, or at least the large blocks from which these are made. The assembly of prefabricated homes from components made by 3D printers is already taking place: the first such house was recently printed in Yaroslavl on a Russian-made printer.[12] Another printer from the same firm, "Spetsavia," has printed an entire office hotel in Denmark.[13]

* * *

Additive technologies already encompass an entire spectrum of technological devices (for extrusion and liquid gas jet pumping, paper lamination, photopolymerization, synthesis of products from powders, direct energy release at specific points) and use all kinds of diverse materials (plastics and plastic products, metals, composites, hybrid materials, materials used in metal casting processes, ceramics, and more).[14]

* * *

3D printing technology is already being combined with biotechnology to create 3D-printed human organs for transplantation. At present, bioprosthetic devices (implants) made from artificial materials are only being used to replace bone and cartilaginous tissue, and for prosthetic hand bones. Attempts at cultivating tissue for human organs (liver, kidneys, bladder, skin) have only really been used so far for testing pharmaceuticals;[15] that said, there is no doubt that the future belongs to these technologies.

12 Rushabh Haria, "Europe's First 3D Printed Pre-Fab House Completed by AMT-SPETSAVIA," 3D Printing Industry: The Authority on Additive Manufacturing, October 25, 2017, https://3dprintingindustry.com/news/europes-first-3d-printed-pre-fab-house-completed-russias-amt-spetsavia-123245/.

13 "The Construction of Europe's First 3D Printed Building Has Begun and Is Almost Complete," 3D Printhuset, accessed October 15, 2022, https://3dprinthuset.dk/europes-first- 3d-printed-building/.

14 For an overview of additive technologies see Aleksandr Prosvirinov, "New Technological Revolution Is Sweeping Past Us," PRoAtom, December 11, 2012, http://proatom.ru/modules.php?name=News&file=article& sid=4189.

15 "Bioprinting Organs on a 3D Printer, How Does It Work?" MAKE-3D, April 12, 2015, https://make-3d.ru/articles/biopechat-organov-na-3d-printere/.

2.6. The Role of IT and Cognitive Technologies

Humanity, as it develops, moves along a path of re*cognition* [*osoznanie*] of its ever-growing needs and re*cognition* [*osoznanie*] of how to satisfy them. This added *cognition* [*znanie*], which has no inherent limits, always reveals not only the desired answer, but a broader horizon of newly forming needs.[16] This horizon is limited only by humanity's ability to recognize and comprehend it at each progressive stage of understanding. This is the essence of human development, scientific and technical progress, and the development of societal relations.

* * *

Thus, at a certain stage, humanity recognized and understood mechanical forces, and included them in its production turnover. Later, this happened with the far more knowledge-intensive force of electricity. By now, our technical and productive base includes informational and cognitive resources.

* * *

Our grasp of these resources is only possible due to computerized digital oversight of production processes themselves, which requires extensive use of informational and communicative networks. This is not the same thing as "digitalization" superimposed on traditional technological processes from the fifth or fourth paradigm. For example, if you removed the numerical control unit from a CNC machine, you would end up with a traditional metalworking machine. But if you tried the same thing with a 3D printer, you would end up with an inoperative unit. If you tried to disconnect "Industry 4.0" from the Web, you would stop entire industries in their tracks.

* * *

The modern tendency towards "digitalization" of the economy might be effectuated even outside the sixth technological paradigm. But the sixth paradigm is what makes this tendency not just economically well-grounded, but technologically predetermined. Without the use of information technology and digital communication, NBIC-convergence would be impossible.

> *Digitalization* is a figurative expression that encompasses a network of solutions linked to the use of modern IT

---

16 *Znanie*, here translated as "cognition" to preserve the author's wordplay, more literally means "knowledge." *Osoznanie* can be defined as "realization" or "comprehension."—*Translator's note.*

and communications technologies (the internet, mobile communications, large-scale information processing, artificial intelligence, etc.) primarily in digital form.

* * *

By using self-learning artificial intelligence (AI) systems, cognitive technologies within the framework of the sixth paradigm encroach upon areas where there was previously no alternative to using human labor. AI systems are already capable of finding, accumulating, sorting, and comparing information in order to make a decision. *It is cognitive technology that, by drawing on the achievements of biotechnology, IT, and communications technology, makes it possible for human beings to directly act upon non-human technological processes* (see: human-machine interfaces, human-machine systems, and human-machine networks).[17] This serves as a new impulse for the production of robotic technology, which is becoming ever more agile, adaptable, and productive.

* * *

AI is still quite far from being able to uncover new knowledge (it can *make use of* knowledge by accumulating and analyzing available data; it can *communicate knowledge through information and communications technology;* but by itself, it *cannot* act as a "discoverer" of knowledge). For this reason, the new technological paradigm places new and growing demands on human beings' research and cognitive abilities. Consequently, approaches based on technological convergence require scholarly interdisciplinarity. An orientation towards technological convergence should be matched by convergence in education. At present, this is significantly hindered by the departmental and sectoral organization of both education and scholarship.

* * *

But why is it that these particular trajectories of technological progress make up the new technological paradigm? And what accounts for the transition from the previously observable "coexistence" and interaction of various technologies to their convergence—that is, to the formation of hybrid technologies?

* * *

---

17 For an overview, see: Milena Tsvetkova, et. al, "Understanding Human-Machine Networks: A Cross-Disciplinary Survey," *ACM Computing Surveys* 50, no. 1 (January 2018), 1–35.

To answer these questions, we must above all pay attention to modern information technology and the related process of technological "digitalization." Information and communications technology, unlike all other technologies, can be diffused into any other technological process. Therefore, "digitalization" is becoming a technological platform capable of uniting multivarious technologies into hybrid technological processes. "Information technology has become a kind of 'hoop' that encompasses all other science and technology."[18] For this reason, digital information technology will be the core of the new technological paradigm.

* * *

Other technologies within this new paradigm are united both by their suitability for convergence with one another and by the fact that this convergence realizes two basic tendencies that characterize the current stage of technological development. These are, firstly, the tendency for human beings to be displaced from the direct process of material production, and secondly, the tendency toward sharp growth in the knowledge-intensiveness of products, and the corresponding relative reduction in their material cost.

---

18 Kovalchuk, "Convergence of Science and Technology—a Breakthrough in the Future," 14.

Step Three

To the Threshold of Technological Revolution

3.1. The Growing Knowledge-Intensiveness of Production

The world is entering the epoch of the next—fourth—industrial revolution, along with a new technological paradigm. Obviously, in the future, *competitive economies* will be those that manage to take the lead not in obtaining and selling natural resources, but in developing and applying *advanced technology*, and which succeed in supplying enough *human capital capable of realizing this technology*. The economic leaders of the future will be technological leaders.

* * *

Transitioning to new stages of technological progress requires us to command newer and newer types of knowledge and to find ways to apply them. The most advanced technologies will also be the most knowledge-intensive.

* * *

We will now turn to changes in *technology*, particularly those which have already become (or are becoming) realities and are occurring within the sphere of material production. First among these, as "post-industrial" theorists have correctly identified, is the growing significance of information technology.

However, unlike the post-industrialists, we do not consider this to be evidence that the decisive role of material production is withering away. We instead draw a different conclusion from this fact: namely, that the *knowledge-intensiveness of material production* is constantly growing.

* * *

By this, we do not mean to merely observe the growing role of information, as many theorists of the information society do:[1] we have in mind less the *production of information* than a new type of *material production*.[2] This difference is fundamental. As the modern world economy shows in practice, the creation of information often leads to the production of "information noise": economic resources are used to create signs[3] and simulacra[4] of useful goods rather than facilitating the growth of labor productivity, qualitative human progress, or the solution of social and economic problems. In the final analysis, this kind of "informatization" leads to the virtualization of social being, which destroys the human personality, its spiritual world, and its social ties.

* * *

Knowledge-intensiveness of the technology of material production is a process that critically synthesizes the achievements of the industrial economy and the information economy. To be precise, this critical synthesis is expressed by the fact that the decisive role in high-tech production is increasingly played by operations and processes in which human beings act not as appendages to machines (tools or assembly lines), but as bearers of knowledge converted into technology: the human being "steps to the side of the production process" and

---

1   "The information society," or "knowledge-based society," has long interested post-industrial thinkers. See Peter F. Drucker, *The Age of Discontinuity: Guidelines to Our Changing Society* (New York: Harper and Row, 1969); Fritz Machlup, *The Production and Distribution of Knowledge in the United States* (Princeton: Princeton University Press, 1962); Yoneji Masuda, *The Information Society as Post-Industrial Society* (Washington, DC: World Future Society, 1981), and more.

2   The issue of industry's knowledge-intensiveness has been debated for quite a long time, but without sufficiently defining concepts like "knowledge-based economy" and "knowledge-intensive industry." See Keith Smith, "What is the 'Knowledge Economy'? Knowledge Intensity and Distributed Knowledge Bases," *United Nations University Institute for New Technologies Discussion Paper Series* #20026 (June 2002), 79.

3   Jean Baudrillard, *For a Critique of the Political Economy of the Sign*, trans. Charles Levin (Candor, NY: Telos Press, 1981).

4   A.V. Buzgalin and A.I. Kolganov, "The Market of Simulacra: a Look Through the Prism of Classical Political Economy," *Philosophy of Economy*, 2–3 (2012).

"comes to relate . . . as watchman and regulator to the production process itself."[5]
In these circumstances, we may speak of the *knowledge-intensiveness* of material
production and the *knowledge-density* of its products.

* * *

The core features of the emerging *new type of material production—knowledge-
intensive production*—are:

- the uninterrupted growth of production's information component and
  reduction of its material component; miniaturization, the tendency
  to reduce the energy density, material density, and capital density of
  productive output;
- new specific features of the *production process* and tendencies in the
  development of *technology* (flexibility, modularity, standardization,
  etc.);
- *network models and networked structures*, replacing vertically integrated
  structures;
- the use of modern methods of management and productive organization
  (just-in-time production, lean production, and more);[6]
- environmental friendliness and an orientation towards *new energy
  sources*;
- the development of qualitatively new technology within material
  production itself, as well as transportation and logistics (nanotechnology,
  3D printing, etc.);
- reduction of the role of traditional manufacturing due to the expansion
  of additive technologies;
- an emphasis on quality and efficiency.

* * *

The application of new knowledge in production is a constantly accelerating
process, conditioned by a growing synergy of useful effects (this synergy is

---

5 Marx, *Grundrisse*, trans. Nicolaus, 705.
6 For more details, see: Taiichi Ohno and Setsuo Mito, *Just-In-Time for Today and Tomorrow*
   (New York: Productivity Press, 1988); William H. Waddell and Norman Bodek, *Rebirth
   of American Industry* (Vancouver: PCS Press, 2005); Behnam Malakooti, *Operations and
   Production Systems with Multiple Objectives* (New York: Wiley, 2013); Sandra Tillema and
   Martijn van der Steen, "Co-Existing Concepts of Management Control: The Containment of
   Tensions Due to the Implementation of Lean Production," *Management Accounting Research*,
   27 (June 2015).

inherent in knowledge as a phenomenon). Consequently, knowledge-intensive production enables quicker satisfaction of increasing needs. The heightened level of new technology is accompanied by reduction in the capital, material, and energy density of production, which ultimately opens the possibility of lowering the specific consumption of resources needed to satisfy a standard unit of human need.

* * *

At some point, then, the "knowledge part" of many products begins to substantially exceed the "material part." This conclusion is illustrated nicely by the graph below, with intersecting curves depicting the relative share of material and intellectual input in overall production costs (fig. 1).[7]

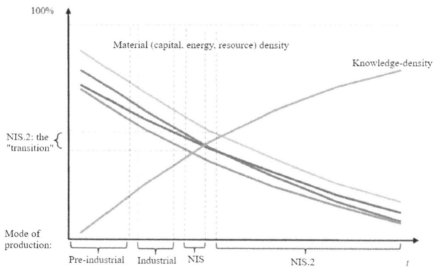

FIGURE 1. The process of historical change in the ratio of the different components of products[8]

* * *

---

7 During panel discussions in the Department of Social Sciences of the Russian Academy of Sciences, this graph was dubbed the "Bodrunov cross." (See R.S. Grinberg, Smart Factories Need Smart People and Smart Economy, *Economic Revival of Russia* 4.50 (2016), 155.)

8 S.D. Bodrunov, New Industrial Society. Production. Economy. Institutes, *Economic Revival of Russia*, 2, No. 48 (2016), 11 (Fig. 4.) *Translator's note:* By "NIS" the author means new industrial society, a.k.a. John Kenneth Galbraith's "New Industrial State," while NIS.2, as will be discussed later in this book, refers to the second generation of the new industrial society.

We can now see that in many ways, this moment (of knowledge-intensity exceeding material density) has already arrived. For example, take an object such as an iPhone. According to information provided by Apple, the material part of this phone constitutes only 4.8% of its overall production cost. Most high-tech industrial products are characterized by similar relative ratios of material input to knowledge, clearly signaling the emergence of the tendency we are describing.

* * *

The development of this tendency provokes a reduction in demand for resources; as a result, the position of resource-extracting countries within the global economy is changing. In terms of the global balance of natural resources, this means less pressure on natural stockpiles, and the possibility to develop in a way that is compatible with preserving (and restoring) an equilibrium with our natural environment.

* * *

New production is knowledge-intensive production, which draws from knowledge-dense technology to guarantee the output of knowledge-dense industrial products that allow growing human needs to be satisfied—including, unlike the mass production of generic products by the first generation of industry, individualized products for particular consumers. This kind of production cannot be supported without a high level of knowledge-density for all its components: materials, labor, organization of the production process, and (we emphasize) the technologies used. What now comes to the forefront—where it will always remain from now on!—is knowledge, in its overt and "pure" form, as the foundational resource of industrial, technological, and societal development.

3.2. Industry 4.0, Smart Factories, and the Displacement of Human Beings from Direct Production

The fifth technological paradigm's development, and the sixth's even more so, has revealed a manifest tendency—already predicted by Karl Marx—for human beings to be displaced from the process of direct production. The automated production lines which appeared even within the framework of the third and fourth technological paradigms, and which also displaced people from direct production, had a narrowly specialized character and could not be recalibrated to resolve new technological problems (such as executing new technological operations and producing new types of products). For this reason, they were only effective when used within very narrow technological niches, mainly the production of raw materials and supplies (for example, chemical production, sheet metal rolling mills, paper-making machines, etc.)

* * *

The fifth and sixth technological paradigms have enabled the production of automated assembly units which can be recalibrated towards different types of productive output and different technological operations. Some of the first visible manifestations of this tendency can be found in universal industrial manipulators (robots) and production machines equipped with computer numerical control (CNC). At this point, the world produces almost no metal lathes without CNC. Meanwhile, the production of robots, which experienced a vigorous start in the 1970s and slowed down somewhat in the 1990s, has resurged, along with progress in microelectronics and artificial intelligence systems.

* * *

The world's leading countries are giving top priority to the development of robotics. The American and Japanese governments have created special organs for robotechnical development. These are the National Robotics Initiative (NRI) in the USA, founded in 2011, and the Robot Revolution Realization Council in Japan, created in 2015.[9] Japan, which has long led the world in its overall quantity and production volume of industrial robots, is faced with the task of maintaining this advanced position. China, which has greatly accelerated its production of industrial robots, is Japan's challenger (during 2018 there were installed 154,000 industrial robots in China, more than in the USA and Europe put together). Currently more than 400,000 industrial robots are produced per year around the world. In Singapore and South Korea, the quantity of these robots already amounts to 7 to 8 percent of the overall industrial employment rate.

* * *

Modern material production has moved far beyond the "factory system"—established in the nineteenth century and surviving up to the twenty-first—in which human beings act as appendages to machines. Already emergent, in our

---

9   Gabor Sziebig and Peter Korondi, "Effect of Robot Revolution Initiative in Europe—Cooperation Possibilities for Japan and Europe," *IFAC-PapersOnLine* 48.19 (2015), 160.

view, are "Industry 4.0"[10] and "smart factories," working in tandem with the "internet of things," or rather, the industrial internet of things,[11] enabling both autonomous technical devices' interaction with one another and human control over them. To facilitate this control, integrated sensors are gaining widespread use, along with systems for processing the large amounts of information received from these sensors (big data). Here we can see the prototype of another version of production—machine-based, industrial, but now "human-free" or "uninhabited" by people.

* * *

Current technologies allow us to automatize not just production processes, but the organization of production, as well as practically the entire cycle of commodity production. Marketing research, which determines the structure and volume of productive output, can be automatized using artificial intelligence. The process of direct production in "smart factories" is based on automated machinery that is coordinated and controlled using "the internet of things." Logistical functions, industrial cooperation between firms, and other forms of business-to-business interaction are also being automatized. Even when interacting with consumers during the stage of ordering, shipping, and delivering products, more and more key functions are being entrusted to artificial intelligence systems. Factually speaking, the only tasks that still belong to human beings are those of development, adjustment, and setting goals.

* * *

In this way, the growth in the role of human knowledge and human reason is accompanied by a gradual displacement of human beings from direct participation in the process of material production. A modern type of "smart industry" is being created, in which *the dramatically increasing role of human reason is paired with the displacement of human beings from direct participation in technological processes.* "Industry 4.0," based on interaction with the "internet of things," is becoming a prototype of this type of human-free production, which nevertheless still relies on the power of human intellect.

* * *

---

10 Germany Trade & Invest, *INDUSTRIE 4.0—Smart Manufacturing for the Future* (July 2014).
11 Hugh Boyes, Bil Hallaq, Joe Cunningham, and Tim Watson, "The industrial internet of things (IIoT): An analysis framework," *Computers in Industry*, vol. 101 (October 2018): 1–12.

These tendencies require us to make sure that as the displacement of people from direct material production snowballs, it does not give rise to a mass of "deprived people," for whom no new jobs, nor adequately dignified living conditions for a certain time period, have been created. In one way or another, the development of production will ultimately provide both these new jobs and these new living conditions. But we must, for example, prevent the emergence of a rift between the decline of obsolete professions and job growth in new spheres, or of an interim period, lasting many years or even decades, filled with millions of unattached "new tramps" and "new beggars" who either live on public handouts or face oppression and harassment.

3.3. The Integration of Production, Science, and Education

In the long term, the next ten years will see the world's leading countries transition to a new technological paradigm in which continuous technological transformation will become an indispensable part of the production process. This accounts for new demands to integrate production, science, and education. We will have to constantly reeducate ourselves.

* * *

Neither reindustrialization, which has now become a vital imperative, nor the further successful development of modern production (not to mention future industry) are possible without *profound integration* of production with education and science—both in ideological and practical terms. The **integration of science, production, and education** is a *necessary organizational condition* and a *pre-requisite for the practical realization* of reindustrializing the Russian economy, which has suffered particularly badly from deindustrialization.

> The *integration of production, science, and education* means the re-formation of these spheres into an interactive system on a national level, including the links in the production chain where production is organically joined with research and development and the continuous process of training and retraining personnel.

* * *

The integration of production, science, and education is a powerful developmental trend of modern global industry. For this reason, the integration of production, science, and education is one of the main priorities for state economic regulators in many leading industrially developed countries.

* * *

Even in the mid-1980s and early 1990s, steps were taken in the USA and in industrially developed European countries to research and implement projects to create and strengthen systems for technological cooperation between industry and the academic community. Beginning in the mid-1990s, Japan also adopted several legislative acts to facilitate the establishment and strengthening of ties between the private sector, academia, and the state. For many years, techno-scientific cooperation between industry, academics, and the state has been the strategic orientation of Japan's national innovation policy.

* * *

An industrial and technological revolution awaits us, and its leaders will be those who manage to ride the "ninth wave" of technological transformation.

This has important practical implications. We must work to bring the components of the new industrial process as close together as possible, shortening the path that must be traveled to implement knowledge in products, skills, and competencies. In other words, we must implement what we call the integration of production, science, and education, creating industrial aggregates that will serve as constituent parts of a new type of industrial sector, and that will, in the future, replace present-day production facilities of the traditional type.

* * *

Technological progress is impossible without scientific development. We must not forget that science is not merely a miner, sublimator, or processor of knowledge. Science's leading role manifests itself only when it is applied to material production, in which it becomes a conductor of knowledge to the technological process and, ultimately, to industrial products. Thus, the tempo and efficacy with which scientific knowledge is transferred to industrial production is of utmost importance. From this follows the need for science, production, and education to be as closely integrated as possible, which we have already emphasized.[12]

* * *

---

12 See S.D. Bodrunov, "Integration of Production, Science and Education as the Basis for the Reindustrialization of the Russian Economy," *Economic Revival of Russia*, 1 (2015).

Because the continuity of innovative processes must be ensured, business subdivisions specializing in research and development are included into the main units of production. The need for R&D assumes constant renewal of the knowledge base used for production, meaning continuous education to retrain staff and raise their qualifications. Learning becomes a ceaseless activity that lasts a lifetime.

* * *

The growing role of knowledge-dense technologies and their corresponding productive resources and results and the need to accelerate the rate at which these technologies are developed and improved are conditioned by macrostructural economic changes. The classical industrial system, which was absolutely dominated by industrial production, and "service society," in which service sectors displace material production, are being replaced by the second generation of the new industrial economy. The dominant position in this new economy must belong to the industries with the most knowledge-dense output—including both those that produce identifiable products and those that create knowledge itself and give people the capacity to acquire this knowledge and apply it to material production.

* * *

In this way, the basis of the twenty-first century's economy should be a composite (see fig. 2) that combines, in its micro- and macro-levels:

- *high-tech material production* that creates knowledge-intensive products;
- *science* that creates know-how;
- education and culture that cultivate people who can acquire necessary knowledge and use it for productive purposes.

* * *

This is how the three main spheres of society's new form of production, whose basis is material production itself, will be formed.

Macro-level

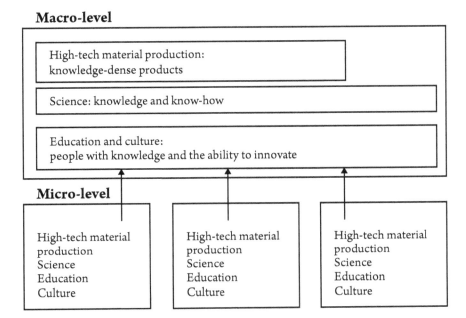

Micro-level

FIGURE 2. The structure of the twenty-first-century economy

3.4. New Ways to Satisfy Needs

The transition to cutting-edge technology changes not only the character of production processes but also the entire economic structure: we see massive shifts in the structure of employment; the structure of human needs noticeably evolves, and along with it, the motivation of human activity. These are not changes that will occur in some unknown future, but changes that are already happening today. Using the capacities of information technology, we can integrate various industrial technologies (mechanical, physical, chemical, biological, etc.) and combine them to solve ever more complex problems and satisfy ever more diverse demands. But is the current world economy up to this task?

* * *

Many of those who consult global statistics opine that the world economy has slowed down continuously over the last two decades, except for certain regions such as China, whose development has until recently relied less on intensive growth than the extensification of industrial stock. But in terms of the satisfaction of human needs, the situation looks entirely different, notwithstanding the perspective of traditional statistics (which utterly fails to meet the demands of real research). In fact, it may be the case that, so far as satisfying its needs is concerned, humanity is currently entering a "golden age."

* * *

Let us pick and examine some use-value [*potrebitelnaya stoimost*] that is designed to satisfy concrete human demands. Take, for example, clocks. They satisfy a need: they tell time. For the sake of argument, we'll assume that twenty years ago, one clock cost one hundred dollars. Around the same time, mobile phones began to circulate. Suppose that the first mobile phones cost one thousand dollars each. A person who bought one of these phones satisfied his need to communicate with other mobile subscribers. Thus, anyone who satisfied both needs at once (telling time and mobile communication) created a demand worth one thousand one hundred dollars (for the clock and for the phone). However, technological development led to technological synergy, which further ensured **synergy in the satisfaction of needs.**

> *Synergy in satisfying needs* is the result of the synergistic interaction of multiple technologies in a single appliance, which allows for several types of need to be satisfied at once.

* * *

After a while, new gadgets appeared that contained two functions—both telling time and linking to the mobile network—while further development made it possible to lower the production cost of a "united" product that met both these needs simultaneously. Let's suppose that this gadget's initial cost was three hundred dollars. Thus, anyone who wanted to satisfy both needs now created a demand worth three hundred dollars. In other words, from the statistical perspective of the global economy, we are dealing with a "collapse of demand," which has fallen from one thousand one hundred dollars to three hundred dollars.

* * *

As measured by standard statistical methods, this would lead to a decline in GDP (see fig. 3.) Here it might be objected: surely the quantity of people who would want to satisfy both needs mentioned above for three hundred dollars is significantly larger than the quantity who might do it for one thousand one hundred dollars. No doubt this is true—the number of people who allow themselves to satisfy two needs for three hundred dollars is much higher than the number of people who let themselves do the same thing for one thousand one hundred dollars. But nonetheless, since there is a physical limit to the number of possible consumers, sooner or later the development of this tendency will statistically lead to falling volume indicators.

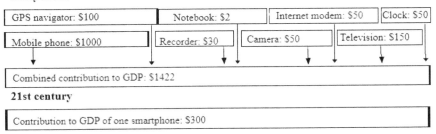

Early 1990s

| GPS navigator: $100 | Notebook: $2 | Internet modem: $50 | Clock: $50 |
| Mobile phone: $1000 | Recorder: $30 | Camera: $50 | Television: $150 |

Combined contribution to GDP: $1422

21st century

Contribution to GDP of one smartphone: $300

FIGURE 3. Synergy in satisfying needs through a single gadget lowers GDP (listed prices are artificial)

* * *

Consequently, we see a divergence in principle between "accounting" and reality, the latter reflecting the factual satisfaction of needs. If we consider the enormous quantity of integrated functions, allowing for ever-growing needs to be satisfied, that are united in new knowledge-dense products, then we will conclude not that economic growth is slowing down but rather the opposite, that (in terms of the satisfaction of human needs) we are witnessing a dramatic rise. One could say that at present, and unbeknownst to ourselves, we are entering an era that will continue to be defined by the progressively more complete satisfaction of people's constantly growing needs, thanks to technological progress.

* * *

In sum, knowledge-intensive products are evolving in terms of their growing ability to appease the ever more far-reaching specter of human need (for example, the evolution from the clock and the phone to smartphones with a colossally expanded range of functions). Technological progress makes it possible to satisfy a multiplicity of human needs with a single knowledge-dense industrial product rather than several simpler ones. Technologies created to meet certain needs simultaneously reveal that they can satisfy others as well.

* * *

Additionally, it is obvious that meeting "old" needs through the knowledge-intensive production of NIS.2 requires substantially lesser expenditure of material resources, even as the share of knowledge in knowledge-dense output is preserved or increased. While each manufactured article requires a certain portion of material, the value of these articles' "share of knowledge" (which may be high at first) is collectively "distributed" among them all. This relative decrease in the resource-density of production, which occurs simultaneously with its increase in knowledge-density, creates a platform for overcoming ecological problems and regulating production's equilibrium with the natural environment.

Step Four

The New Industrial Society's Second Generation[1]

4.1. Technological Shifts Change All Aspects of Production and the Social Order

Shifts in technology give rise to changes in all aspects of the production process, which in turn signal transformations in society's relations of production and in the overall nature of society. Galbraith's new industrial state has receded, along with the unsuccessful attempt at a "post-industrial future." The second generation of the new industrial state is emerging, with the task of resolving the previous epoch's contradictions.

* * *

This is why we need an outlook on the future that has a wider historical horizon, in order to find paths of development that would combine a technician's rational approach with spiritual wisdom in posing tasks and goals. Production

---

1 The author's phrase "novoe industrial'noe obshchestvo" is mostly translated here as "new industrial society." However, readers should be aware that this exact phrase is also the accepted translation into Russian of John Kenneth Galbraith's concept of the "new industrial state" (from Galbraith's 1967 book with this title). Bodrunov's reference to Galbraith is conscious, hence his emphasis on the "second generation," Galbraith's being the first. For clarity's sake, the translation opts for "society" rather than "state."—*Translator's note.*

should not be subordinate to the pursuit of consumption, or prestige, or capital accumulation; it should instead be controlled by human reason. But for this, human reason itself must undergo an evolution that would change the present day's hierarchy of values.

* * *

The technological development of material production is bringing us to a frontier at which humanity may truly begin, for the first time, to free itself from the work of production and from directly acquiring our "daily bread." At the same time, material production, while qualitatively changing, largely retains its industrial character in a technological sense: it is still machine production. The essential difference between the old industrial system and the new one consists in the intellectualization of production, at the level of the knowledge-intensity of production and the knowledge-density of its product.

* * *

This "new industrial" mode of production is so knowledge-intensive that material inputs and the input of human labor are relegated to the background, allowing human beings to practically give up using their own physical strength in the production process while nonetheless remaining "within" production, participating in it, and therefore fulfilling tasks of labor (which are more and more intellectualized). In the process, the extent to which technologies are intellectualized in the "new industrial" mode of production will at last allow humanity to begin its escape from the bounds of production.

* * *

The twenty-first century's new industrial state and economy must become the "negation of the negation," the dialectical overcoming of both the late industrial system described by John Kenneth Galbraith and the post-industrial informational trends investigated by Daniel Bell and his followers.[2] How should this "negation of the negation" be conceived? We propose not to get carried away making pretty utopias, but instead to thoroughly analyze the real trends of modern material production's renaissance.

---

2   The author's language here is Hegelian. "Overcoming," *sniatie*, is Hegel's *Aufhebung*, sometimes translated into English as "sublation." This term refers to the simultaneous cancelation and preservation of the term to which it is applied. Thus, the author's point is that while Galbraith's and Bell's systems are negated, their features are also maintained at a higher level within the second generation of new industrial society. —*Translator's note.*

The new industrial society's second generation (NIS.2) is a society founded on a new developmental cycle of industrial material production, characterized by growing knowledge-intensity, the transition to producing knowledge-dense products, accelerating technological change, movement towards a continuous stream of innovations, and the completion of the process of integrating production with science and education (including at basic stages of production itself).

4.2. Characteristic Features of NIS.2

What we need is not just a "technological leap," but advancement in *all aspects* of modern material production (materials, labor, production and the application of knowledge, organization of production). Only then will it be possible to speak of our emergence into the next generation of new industrial society—NIS.2. This is why Russia, whose national economy has been undermined by an unprecedented 25 years of deindustrialization throughout the post-Soviet period, requires economic reindustrialization on a new high-tech basis, as we have discussed in numerous publications.[3]

* * *

At the new stage of industrial society's development (NIS.2), the tendency for technological change to accelerate takes on principal importance. "Accelerating acceleration" is becoming the distinctive sign of the coming society's economic system. The most important thing now is the rate at which scientific discoveries and achievements are transmitted to industrial production and its components, particularly products: industrial production is acquiring a constantly innovative character. Such an aspect of innovative activity as technological *transfer* has now been included within the production process, not as an episodic moment at which an innovation is "introduced," but as an essential and ceaseless element of effective modern productive activity.

---

3   S.D. Bodrunov, *Formation of the Strategy of Reindustrialization of Russia* (St. Petersburg: S. Y. Witte INID, 2013); S.D. Bodrunov and V.N. Lopatin, *Strategy and Policy of Reindustrialization for Innovative Development of Russia* (Moscow, St. Petersburg: S. Y. Witte INID, 2014); S.D. Bodrunov, *Formation of the Strategy of Reindustrialization of Russia*, 2nd revised and expanded edition (St. Petersburg: S. Y. Witte INID, 2015); S.D. Bodrunov, ed., *Integration of Production, Science and Education and Reindustrialization of the Russian Economy. Proceedings of the International Congress "Revival of Production, Science and Education in Russia: Challenges and Solutions"* (Moscow: LENAND, 2015); and more.

* * *

Changes in production's technological basis during the transition to the next generation of new industrial society are inevitably accompanied by **changes to the system of economic relations and institutions**, conditioned by the development of societal production's new content and structure. This economy, assuming that features of the industrial past have been reborn in qualitatively new form, gives rise to new appeals for both the market's self-regulation and private property rights, on the one hand, and state economic intervention, on the other.

* * *

Indeed, the individualization, flexibility, and knowledge-density of production, the widespread use of internet technologies in material production and subsequent trade, the increasing importance of workers' individual professional qualifications—all this energizes the development of small and mid-sized businesses, underscoring the need for broader economic freedom. The personal experience, energy, and talent of each innovator-entrepreneur will be crucial. In this regard, the twenty-first century's new industrial economy is the negation of the negation of the era of classic industrial capitalism and the beginning of late industrial capitalism, which fostered the formation of great industrial empires.

* * *

The twenty-first century's new industrial economy differs from that earlier era in principle. The challenges of the present particularly require the systematic development of many areas of public state management. Among these tasks, far from unimportant is the establishment of the pure and applied sciences as one of the basic branches of modern production, along with the development of universally available public education at both professional and higher levels, given the constantly increasing qualifications of workers.

* * *

The development of complex integrated production units, the macro-economic integration of production, science, and education, the problems of structurally rebuilding modern economies, the challenges of ousting overly developed and entrenched economic intermediaries—all these goals point to the need for active state industrial policies and long-term public-private investment partnerships. These conditions should determine all other aspects of state regulation of the economy.

* * *

Substantial demands on economic relations and institutions have also been made by the transition to the large-scale creation and use of knowledge-dense products. The synthetic nature of these products is the result of many changes in the system of economic relations and institutions. To be precise, ownership of such a product includes a system of rights that encompass both the material object itself and its intellectual component. For many high-tech products, the costs of their technological development and maintaining intellectual property are comparable with their straightforward production costs; in some cases, the former exceed the latter. This indicates the principal importance of questions of intellectual property for the new industrial economy during the transitional period toward NIS.2.

* * *

This society—NIS.2—truly will become new. First and foremost, its socioeconomic relationships will be new. New industry will make it necessary to give a new face to both markets and state regulation, along with private enterprise and state property. Due to the fundamentally different and practically limitless broad availability of means to satisfy non-simulative human needs, under NIS.2 the ownership or appropriation of an individual product will become far less meaningful. Classical Marxist thought maintained that the fundamental contradiction of capitalism was between the social character of its production and its private means of appropriation. Under NIS.2, production becomes "detached" from people, while "appropriation" becomes the simple and extremely accessible satisfaction of need, without harming other individuals.

* * *

This possibility arises with the further technological progress of the industrial mode of production. By developing the next generation of technologies, humanity does not forswear the industrial process, but grounds it upon a controlled and directed natural process.

* * *

Alongside transformations in production's technological foundation, all its other components change as well: labor, products, and the organization of production. *But the main thing that should be emphasized is that all these transformations lead to transformed economic relations, including property relations, in this new generation of industrial society.*

4.3. Shifts in Property Relations and All Other Economic Relations

Even at our current level of social development, before the transition to NIS.2 has taken place, evolutionary trends in property relations can be observed, tending toward their socialization and erosion. Property relations, especially private property, are supposed to secure the proprietor's incontestable rights to ownership, use, and full disposal of economic resources. Yet for some time, the evolution of economic relations has led to the accretion of all sorts of inhibitions on property rights, aimed at guaranteeing the social responsibility of the proprietor.

* * *

Among these, we can point to legal easements on privately owned land, which allow third parties to exert their rights, within certain limits, to use that plot of land (the right to pass and travel through, the right to access water sources, the right to drive cattle, access rights to stretches of coastline, the right to lay communications, etc.) Due to construction, transport, and industrial activity, property rights are already limited and inhibited in multiple ways, relating to the obligation to fulfill security needs, the observation of specific quality standards, ecological needs, and more.

* * *

We should pay especially close attention to the evolution of intellectual property rights, which regulate the economic circulation of modern production's main resource—knowledge. Here we can see phenomena such as crowdsourcing, wikinomics, free software, open source, copyleft, and so on. This all leads to the development of regimes of free access to intellectual resources. On the other hand, a rather bitter struggle is underway to "enclose" intellectual property.

* * *

This corresponds to two tendencies in the development of property relations, both of which can be traced in our current economic system: 1) the conservation of previously established relations, and 2) the erosion of property rights, even to the point of denying them altogether.

* * *

The erosion of property rights manifests itself in the development of forms of shared ownership and use of property, as well as the separation of ownership from use. An owner may temporarily decline to use their property and transfer usage rights to another person: see leasing, coworking, and various types of shared use (car-sharing, kick-sharing, time-sharing, and more). The sharing economy amounts to hundreds of billions of dollars per year. In China alone, the sharing economy reached 1.05 trillion dollars in 2019, nearly 8 percent of GDP.[4]

* * *

To a significant extent, the transition to temporary use of property without obtaining ownership rights is determined by the growing speed of technological change. There is no economic sense to obtaining full ownership of units that will become obsolete in a few years. Often the owners of these units take on additional obligations to repair and modernize them for their users.

* * *

Another tendency that also leads to the erosion of property is the fragmentation of capital. Not for nothing has the contemporary "economic theory of property rights" paid so much attention to the problem of legal authority's disintegration and the erosion of property rights.

* * *

The emergence of stock ownership of corporations already leads to yet more complex fragmentation of property rights rather than to straightforward ownership and use. Stockholders do not hold property rights over the full volume of capital. Moreover, the totality of their legal entitlements depends on the type of stock and the volume of the block of shares.

* * *

The functions of appropriation within the framework of property relations have also undergone a major evolution: even in the first half of the twentieth century it became obvious that these functions had been split up between the owners and the managers of capital. These problems were well understood by many researchers (Thorstein Veblen, Adolf Berle and Gardiner Means,

---

4 "China sharing economy market to exceed 9 trln yuan: report," Xinhua News Agency, November 2, 2019, http://www.xinhuanet.com/english/2019-11/02/c_138523206.htm.

Stuart Chase, and others)[5] even before James Burnham acquired undeserved renown as their discoverer; Burnham merely dressed them up with the flashy term "managerial revolution" and the assertion that capitalist society would be replaced by managerial society.[6]

* * *

In fact, property's disintegration of functions goes even deeper than the distinction between stockholders and managers. Galbraith demonstrated that real usage of capital increasingly belongs to a whole army of specialists that forms a corporation's "technostructure." But this is not all. The final users of capital's various elements are the corporation's hired employees—though, of course, each of them individually performs only a small and partial function.

* * *

Moreover, in all these cases, there is a certain "stratification" and "splitting up" of property along several lines: 1) appropriation-ownership-management-use; 2) dispensation of each of the elements of the set of property rights among numerous actors in space and/or time and/or 3) by function (stakeholder-manager-worker); 4) in terms of power and authority.

* * *

This last aspect merits some commentary. Power and property are correlative concepts. Property guarantees the proprietor's power over what he or she owns (even to an extreme degree: see, for example, slave owners). As such, a breach of property rights is typically linked to a manifestation of will—that is, either formal-legal or informal, even criminal, violent action. The above-mentioned forms of erosion of property rights lead to step-by-step stratification of power and authority, and also to the overcoming, the "falling asleep," of the political aspects of social relations in the sphere of production in the broad sense (unified with trade, distribution, and consumption).

* * *

From this we conclude that the meaning of power as an institution will be decreased/eroded/fragmented (which is already happening in a historical

5   See Veblen, *The Engineers and the Price System*; Berle and Means, *The Modern Corporation and Private Property*; and Chase, *A New Deal*.
6   James Burnham, *The Managerial Revolution: What Is Happening in the World* (New York: John Day Company, 1941), 71.

sense). Accordingly, the role of the state as the subject of power and authority, as the general *owner* of the right to develop society, will also decrease as we advance towards the noonomy.

* * *

We may also note the direct influence of technological change on property relations. "Blue-collar" and "white-collar" jobs are being replaced by robots and artificial intelligence. What will happen to property relations when the task of performing many economic functions moves from humans to technogenic beings? For example, how can property users be held responsible for an accident caused by a robot driver? Responsibility for harm could then be placed on the owner. But what about responsibility for breaking the rules of the road?

* * *

Little by little, functions of property use and even management are already drifting away from human beings. Further evolution in this direction will only get faster.

These processes, along with the above-mentioned tendency for the worth of property ownership to decrease, will lead to transformations not only in the property system but also in the entire social order. With great certainty, we can predict that *the stage of NIS.2 will be governed by an economy of shared use, an economy of fragmented and eroded property rights.*

* * *

Thus, the system of property relations is substantially changing as we transition to NIS.2, which leads to the transformation of the entire system of economic relations. The market's character is changing: more and more, a greater role is played not by the spontaneous fluctuations of market conditions, but by the results of complex mutual agreements between people who hold diverse and interlocking elements of specific property rights. The character of state regulation is also changing, as it begins to orient itself toward creating a consensus from the complex balance of economic interests that arises from the new nature of property relations and the new modification of market relations.

4.4. The Evolution of Labor and the Evolution of Needs

The emergence of the industrial mode of production saw the development of contradictions in the formation and satisfaction of human needs, as well as in the means to resolve these contradictions. The industrial mode of production is based on the ability to mass-produce standardized products. In turn, this ability prompts the formation of demand for mass consumption. But mass production and mass consumption did not immediately "meet." A series of acute social conflicts were needed, over the course of the nineteenth century and the first half of the twentieth, for mass production to result in mass consumption, at least in the most developed countries.

* * *

Once the ability for mass production to be combined with mass consumption was achieved, it led to an even greater expansion of needs, along with growing ability to satisfy them. Technological application of knowledge made it possible not only to create new objects of consumption and services, nor merely to ramp up their output. At the same time, a decrease occurred in the proportion of material resources used to produce products, and an increase occurred in the proportion of knowledge objectified in them (this was the growth of products' *knowledge-density*). Without this tendency, mass production, spurred on by mass consumption, would have long ago reached the absolute limit of its available resources (though this threat has still not yet been taken off the agenda).

* * *

In recent times, the development of science and technology has given rise to a new tendency: the ability to create items that simultaneously meet several needs. Meanwhile, as the growth in the volume indicators of production and consumption slows down (or even decreases), it becomes possible to raise the level at which needs are met.

* * *

The level at which needs are met turns out also to be inextricable from changes in the character and structure of needs. Usually, these transformations are ascribed to Maslow's so-called pyramid, in which fulfilling the needs of a lower level causes emphasis to switch to the needs of the next-highest level. But the fundamental causes of change in the structure of needs belong to the sphere of production, not consumption.

* * *

The growing knowledge-intensity of production also signifies the growing knowledge-intensity of people's labor activity. The displacement of people from the direct production process, the concentration of human beings' role in management and goal-setting, shifts human activity to a primarily creative arena, connected to the discovery and technological mastery of new knowledge. For people taking on this role, needs associated with personal development, which amount to prerequisites for deploying one's creative powers, will become primary.

* * *

This exact transformation of the content and structure of needs is the most important condition for sating them. When the motivation to develop oneself becomes principally important, the drive to quantitatively ramp up the consumption of material goods is weakened (assuming such consumption has already been ensured to the extent required for normal maintenance of human beings' life needs). This transformation in the character of needs, in turn, serves as a prerequisite and stimulus for the development of creative activity in production.

Step Five

Civilization at a Crossroads

5.1. Rising Technological Opportunities and Rising Risks

To understand the new possibilities that arise thanks to modern technology, attention should be paid to a particular aspect of what happens when technologies are combined. There are many different unique forms of technological synergy that result from such combinations. That said, in our view, these results cannot be described in terms of established theories (for example, the theory of waves) and leave space for further research.

> *Technological synergy* is a phenomenon that amplifies the usual effects of technology, exceeding the sum of the effects of particular technologies by combining two or more technologies.

* * *

Technological synergy, like any introduction of new technologies into an existing technological environment, are possible due to an effect that we have named *penetration*. Of course, new technological solutions cannot be incorporated into all technologies. We designate a technology's receptiveness to penetration using the term *readiness*.

Penetration is the insertion of a new technological solution into other technologies and into various elements of the production processes based on those technologies.

Readiness is the potential for a new technology to be accepted into new technologies and various elements of the production processes based on those technologies.

* * *

It is because of the effects of technological synergy that the hybrid technologies that characterize the sixth technological paradigm can function (as discussed in an earlier chapter). The sixth paradigm's technologies are high in readiness to interact with one another, which is why they can form hybrid technologies on a massive scale. And precisely the effect of one technology's penetration into another is what determines the formation of the sophisticated integrated technical complexes that create technocenoses.[1]

A *technocenosis* is a community of technical instruments (analogous to a biocenosis) characterized by technological interdependence and common aims.

* * *

The potential of new technology is deceptive. Technology dangles before us the prospect of improving the human species, freeing ourselves from the limits of our frail bodies and the biological thinking instrument known as "the brain." Yes, it is tempting to leap beyond these barriers . . . But what for? To expand our ability to satisfy our purely animal instincts to consume as much as possible? What if we made a different choice?

* * *

Both science and society at large are already beginning to understand not only that the sixth technological paradigm may rebuild individual and collective forms of human life in their entirety, but that the new technological paradigm itself, in turn, may only reveal its full potential within a new *social* paradigm. As the President of the Davos World Economic Forum puts it: "The more we think about how to harness the technology revolution, the more we will examine ourselves and the underlying social models that these technologies embody and

---

1 The concept of a "technocenosis" was coined by Boris Ivanovich Kudrin. See B.I. Kudrin, "Studies of Technical Systems as Communities of Technocenosis Pieces," in *System Studies. Methodological Issues. Annals 1980* (Moscow: Nauka, 1981): 236–254.

enable, and the more we will have an opportunity to shape the revolution in a manner that improves the state of the world."[2]

* * *

The technological prerequisites necessary for transitioning to a different means and different level of meeting human needs are presently being created. At the same time, the mechanism by which these needs are formed is itself changing. This, in turn, leads to a mass of changes in social relations and institutions and, finally, in the social conditions that set the direction of technological progress itself.

* * *

Technological forces, once awakened by humanity, will not be stopped without direct and all-encompassing control by human reason, which would in turn undergo changes so that it would be able to set fruitful, rather than destructive, directions of development.

In the most direct sense, changes in the technologies and social relations of the future are linked to the birth of a new type of human activity, which means the creation of a new type of human being.

* * *

Human beings stand at the threshold of one of the most important crossroads in our history:

- we may turn towards a genuinely rational humanity;
- we may, instead, choose a dead-end path towards a technetronic society in which elites satisfy their own (immeasurably expanding and primarily simulative) needs, while the majority of people are employed in the service sector, which increasingly becomes a "servile sector"—potentially accompanied by loss of control over the development of the technosphere and destruction of our habitat.

* * *

Technological progress brings not only potentially positive prospects, but also—without humanity's due attention to the risks of "improperly" using the results of progress—substantial dangers. In this sense, we now witness

---

2 Klaus Schwab, *The Fourth Industrial Revolution* (Geneva: World Economic Forum, 2016), 9.

the advanced development of the technosphere, while the part of social consciousness that is "responsible" for making reasonable use of technological achievements, and sustainably forming non-simulative individual and collective needs, lags behind.

* * *

The level of technological development attained by humanity is already extraordinarily high, making it possible to do irreversible harm to civilization— so long as there is no corresponding "balance" of public awareness to put a stop to this scenario. In this sense, the modern state of civilizational development should be characterized as a crisis situation. A great many negative tendencies have already accumulated in the development of the technosphere.

* * *

Humanity's biological habitat is under threat, while problems are amassing in people's relationship to the technosphere, including rising dependence on a high-tech informational environment, leading to a kind of "cyborgification" of human beings (even without the formal intrusion of machinery into physical human bodies). Humans are confronting the increasing precarity of their existence as both biological and social beings.

* * *

It is vital that we forestall the possibility of a wrong choice at modern civilization's developmental crossroads during the transition to NIS.2. Two basic scenarios are likely at this point.

* * *

One of them might tentatively be called the "technocratic" scenario. At the moment, this is the path we are persistently following, with no light visible at the end of the tunnel. This scenario is based on the currently accepted global paradigm of "economic development," in which progress is understood not just qualitatively but quantitatively. In essence this is a savage process, which proceeds from the animalistic side of humankind. Figuratively speaking, its goal is to gobble up as much as we can, even if we burst. And if we do not produce and consume more today than we did yesterday? "Stagnation! Recession! Decline in the level of popular needs met! Impoverishment!"

* * *

But what, in principle, is the important thing about satisfying needs? Quantity? Or quality? If we are talking about non-simulative needs, then the answer is quality (and this quality is what in fact determines the quantitative measurement of needs). But according to the harmoniously balanced statistical number games of today's "funhouse mirror economy," this algebra is incorrect! And unless we repudiate this path, the path that the whole world is fixedly marching upon right now, we will slide downward towards the technocratic version of development. This threatens us with a battle to exhaust the world's resources—and the technocratic model is fully armed with all the latest technology.

* * *

Our current economic system is gradually "germinating" into NIS.2. But this stage of economic and social development has a transitional character. The progress of sixth-generation technologies inexorably presents us with a choice: either humanity will remain the same while changing its technological and socioeconomic system; or the system will change humanity; or both will change. Obviously, both tendencies will exert themselves to some degree. But which will prevail? Humanity, with its principles of communication and self-development? In that case, the production of our material conditions of existence will remain at the mercy of technetic beings (sprouting from the coming "Industry 4.0," from artificial intelligence systems, etc.)

> *Technetic* means related to technical and technological reality.

> *Technetics* is the science of technical reality.

* * *

It will not be humans that occupy themselves with the needs that can be met using technology. But the determination of "technical tasks," the setting of goals, will remain up to humanity. Yet the goals set for production directly depend on society's predominant values. This means that values, too, must change to meet the task at hand. Any mistake in formulating the goals of such a highly developed technosphere, especially one with relative autonomy from human beings, will be extremely costly. If the goals of this kind of production are defined using the old prevailing value system of today, acute contradictions will inevitably emerge—both socially and in conflict with our natural environment.

* * *

Let us imagine that at a certain point, "quantitative" movement along our current direction of development crosses a qualitative boundary, causing an explosion—and then the birth of a new civilization . . . What would it be like? Civilization can develop in two ways. The first option is a technetronic civilization, that is, the *de facto* eradication of humanity as we currently know it and the appearance of other beings, more capable of existing in that environment. The second option is that humanity can knowingly, consciously become the creator of a different path, which we have dubbed "noocivilization."

* * *

There is a simple way to realize the first option: we just have to maintain our current predatory course, "developing" our present "economy," creating new simulative needs, and satisfying them by attaining newer and newer (technetic, technogenetic) products. In other words, we are traveling along a path of technological genesis, and eventually these new products will create a new environment all by themselves.

* * *

Needless to say, scholars who are expanding the horizons of scientific knowledge are driven by good intentions—the creation of new types of medicine, for example, or the correction of genetic abnormalities. But they also do not deny that their scientific achievements could be used to create new forms of life and "edit" humanity's very biological essence. How far will we travel down this path, and what criteria will guide us as we make the decision? Our choice at the crossroads of present-day civilization will depend on how we answer these and many other analogous questions.

5.2. Satisfying Needs: Reasonable or Simulative?

As production constantly expands its technological abilities with respect to meeting constantly growing needs, what precise human needs will orient it? And how will these needs be shaped?

* * *

Capital always pursues the expansion of mass production and mass marketing. On one hand, this impulse gives rise to the constant development of production, the improvement of technology, the progress of productive forces; at the same time, it expands human needs and makes them more diverse. But since economic rationality is indifferent to the nature of specific needs and the means used to satisfy them---all that matters is that they attract the consumer's effective demand—the progress of production and consumption develops alongside industries that formulate and satisfy artificially imposed needs, playing upon human weakness.

* * *

But the modern marketplace not only plays upon human weaknesses in order to indulge them by expanding production and marketing. The market creates false, illusory, *simulative needs*, along with the corresponding means to satisfy them, which in turn can only simulate the satisfaction even of illusory needs.

> *Simulative needs* are illusory, false needs, imposed by the market system in its single-minded pursuit of greater sales.

* * *

Thus, the market economy increasingly becomes less a space in which real use-values are produced for the satisfaction of real needs, and more a world for creating *commodity-simulacra* that simulate the satisfaction of simulative needs, artificially created using marketing, advertising, and various forms of manipulation of consumer consciousness, which have become remarkably widespread due to the growing use of information technology.

> *Commodity-simulacra* are goods that signify the satisfaction of simulative needs or means for the imaginary satisfaction of imaginary needs.

* * *

The nature and role of simulative commodities, or simulacra, or just signs of the satisfaction of imaginary needs, was studied in detail by Jean Baudrillard (in his *For a Critique of the Political Economy of the Sign*) from a social and philosophical perspective. But a simulacrum is not only a social phenomenon. Mass production of simulacra has led to the emergence and formation of an extensive market for them, which has become a significant socioeconomic phenomenon.[3]

* * *

---

3 For an analysis of commodity-simulacra and their market, see A.V. Buzgalin and A.I. Kolganov, "The Market of Simulacra: a Look Through the Prism of Classical Political Economy," *Alternatives* No. 2 (2012): 65–91.

The law of the elevation, expansion, or "escalation" of needs also operates in the simulative sphere as a law of the escalation of false, fake needs. This is because once the ability to satisfy some need or other has been realized, another thought always follows: what kind of new need will appear next? This occurs due to the nature of knowledge itself: each "quantum" of knowledge obtained not only responds to the utilitarian question that prompted the search for that "quantum" in the first place, but imparts a broader content, giving rise to new, "supplementary" knowledge. This implies the possibility of formulating new and more expansive needs, and so the old need "germinates" into new needs.

* * *

But when and why do these particular "deceptively necessary" needs emerge? Because, as biological creatures, human beings, as soon as they begin to understand the external world (and themselves as an enduring being within it), start to think about stockpiles, reserves, the future—at least, the next step forward. So, humans try to calculate and evaluate what it is they might end up needing, drawing on their accumulated knowledge about themselves and their potential needs. Thus, once the chance appears to create a stockpile, humans will take it.

* * *

However trivial it may seem, this is the source of all ideologies of accumulation for some purpose or other—for example, the ideology of acquiring extra space, even when it is not really needed in that moment and may later cease to be felt as a need at all! Step by step, our correct desire for potentially necessary things, that is, our desire to accumulate natural goods, begins to cross a certain line, at which point we no longer know exactly how much we need, but we understand that at some moment whatever we have may not be enough.

* * *

However, the natural needs that appear in a world where the conditions of humanity's existence are neither stable nor guaranteed do not take on particularly strict boundaries. Any reserve or volume of resources obtained turns out to be inadequate; under these circumstances, the mere availability of some measure of confidence in the future starts to be measured as the size of one's "mountain" of accumulated goods. This impulse is socially reinforced: the accumulation of wealth becomes a symbol of success, a symbol of a person's social status, and the pursuit of this status is equated with enhancing one's volume of acquired (though perhaps not even really needed) wealth.

* * *

In this way, simulative needs arise as everyday needs are met. But there is already a distinction to be made here: some simulative needs may be satisfied even though it is essentially illusory (meaning that people do not need that much at that moment, nor in the foreseeable future or at any other time). Nonetheless, such needs can be met. Meanwhile, other needs can emerge that amount to a pure simulation of rational needs and that, in principle, *cannot be met* at the given stage of development, but which one can think about. We may call these latter needs *phantasms*, while the first kind of simulative needs can be called *excesses*. (See fig. 4.)

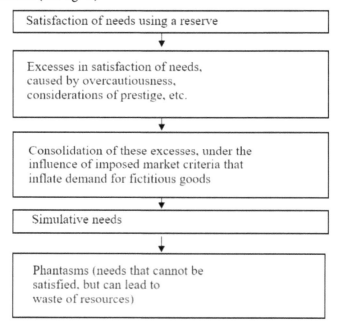

FIGURE 4. The evolution of simulative needs

* * *

Bear in mind that in certain cases, simulative needs may transition with time and become non-simulative and vice versa. Moreover, in the same moment, one person's real need may be another's simulative need. For instance, the desire for a precise-fitting dress or facial skincare products would have been a simulative need for peasant women in past centuries ("it's not about thriving, it's about surviving!"), while nowadays this has become a normative need. Meanwhile, a slide rule, which was once a necessity for every engineer, is probably only needed today by collectors of antique measuring devices.

* * *

These days we are moving—in fact, racing—along a path of escalating satisfaction of an ever-greater quantity of unreasonable needs. In fact, our entire current economic paradigm is tailor-made to do this! More, more, more . . . but right now, are there more non-simulative or simulative needs in this "more" (not to mention phantasms)?

* * *

We must deal with the fact that the modern market economy lends itself to unusual inflation of simulative needs in the pursuit of higher sales. It is no coincidence that the production and consumption of simulacra has become so widespread in recent decades. The deep causes of this phenomenon are the structural shifts in societal production that took place at the turn of the 1980s, when the world was engulfed in postindustrial economic myths. These myths did not arise in a vacuum: *the unrestrained growth of the service sector on one hand, and deindustrialization on the other; the virtualization of everything which fostered all of this—these are the material foundations of the expanding production of commodity-simulacra and dissemination of simulative needs.*

* * *

Simulative need is not just a question of individual personal choice. The inflation of simulative needs is echoed in the real economy by the growing expenditure of real resources on things that only give the illusion of usefulness. And the increased production of commodity-simulacra is one of the substantial factors behind the escalation of the resource burden placed on the natural environment.

5.3. Rising Pressure on the Environment

The concept of noosociety is undeniably connected to Vernadsky's idea of the noosphere. Rationally speaking, his overall conception of the biosphere's transition to the noosphere can hardly be contested. Vernadsky's basic thesis—that humanity was becoming the most important geological force, and that from the twentieth century onward, humans would be responsible for the reproduction of the Earth's biosphere—has been repeatedly validated by historical practice, in both positive and negative ways. *Technogenesis*[4] already

---

4 The term "technogenesis" was coined by the Soviet academic A.E. Fersman: see his *Geochemistry*, vol. 2 (Leningrad: Publ. Academy of Sciences of the USSR, 1934) 27. See also R.K. Balandin, *Geological Activity of Mankind. Technogenesis* (Minsk: High School, 1978). For

rivals biogenesis and the biosphere in terms of the overall mass of substances it involves and its energy costs.[5]

> *Technogenesis* refers to the creation of the technosphere and the process by which it is filled with techno-substances and technetic beings.

* * *

The history of civilization's development shows us that the accelerating growth of the "technetic species" created by humankind (in strict correspondence with the law of innovation's "accelerating acceleration") damages and crowds out biotic diversity. As a result, the escalating environmental burden—related to the growth of people's simulative needs, the use of natural resources to meet those needs, and the expanding footprint left by mining and processing those resources—opens the real possibility of negative (or catastrophic!) consequences. (See fig. 5.)

* * *

In terms of resources, we must unequivocally stake our future on deprioritizing traditional (material, tangible) resources in favor of the basic resource of NIS.2—that is, knowledge, as embodied in technology. But in epistemological or gnoseological terms, we must change the priorities, purpose, and orientation of our development.

* * *

This is well illustrated by data showing the current state of our general civilizational environment, which we have created by following the trends of the present paradigm of "economic growth." There are two major versions of how events might develop from this state of affairs. Take, for example, the overall volume of everything that humanity has created in its five thousand years of existence: geologists estimate that *the mass of the technosphere*, that is, everything that humanity has created throughout its history using technology, *amounts to thirty trillion tons.* (For a more detailed breakdown, see Table 1.)

---

a definition of technogenesis, see B.I. Kudrin, "Technogenesis," in *Global Studies Encyclopedia*, ed. I.I. Mazur and A.N. Chumakov (Moscow: Raduga Publishers, 2003), 998.

5   A great deal of data about technogenetic pressure on the biosphere can be found in Igor Anatolievich Karlovich, Regularities in Technogenesis Development in the Structure of Geographical Shell and Its Geological Consequences (PhD thesis, Vladimir State Pedagogical University, 2004).

Unrestrained, growing, and mostly simulative needs and desires

The possibility of losing control over the technosphere's development, due to its orientation towards pursuing artificially inflated needs and desires

An already exceptionally high level of technological development, making it possible to do irrevocable harm to civilization

Humanity's growing dependence on its technical and informational environment

The accelerating growth of "technetic species" created by humans, which damages and rapidly crowds out biodiversity

Technology's increasing environmental toll

Premature development of the technosphere, while the part of human social consciousness that "takes responsibility" for the rational usage of technological achievements lags behind

Weak internal regulators of rational behavior, as defined by the content of culture and its level of development

FIGURE 5. Factors in the crisis scenario of civilization's development

* * *

TABLE 1. Approximate mass of the major components of the physical technosphere, arranged in order of descending mass (where 1 Tt = 10^{12} metric tons).

Component	Area (10^6 km^2)	Thickness (cm)	Density (g/cm^3)	Mass (Tt)	%
Urban areas	3.70	200	1.50	11.10	36.9
Rural housing	4.20	100	1.50	6.30	20.9
Pasture	33.50	10	1.50	5.03	16.7
Cropland	16.70	15	1.50	3.76	12.5
Trawled sea floor	15.00	10	1.50	2.25	7.5
Land use and eroded soil	5.30	10	1.50	0.80	2.7
Rural roads	0.50	50	1.50	0.38	1.3
Plantation forest	2.70	10	1.00	0.27	0.9
Reservoirs	0.20	100	1.00	0.20	0.7
Railways	0.03	50	1.50	0.02	0.1
Totals (where applicable)	81.83			30.11	

Source: Jan Zalasiewicz et al., "Scale and diversity of the physical technosphere: A geological perspective," The Anthropocene Review 4, no. 1 (April 2017): 12.

* * *

By now, humanity has already devoured so much dead nature, so much of the mineral world, that it has created much more over the past five hundred years than nature (non-biological "civilization") had "devoured" over the course of hundreds of millions of years. According to other geologists, this means that we may speak of the arrival of a new geological era. Experts call this new era the "Anthropocene."[6] But geologists describe this era from an external perspective; by contrast, the author's approach is based on what the era is like "from the inside," what it "grows" from—namely, our unintelligent (or intelligently unintelligent) use of technology.

* * *

Here is another estimate: according to biological data, *the weight of the Earth's cumulative biomass throughout its 4.5 billion years of existence*—that is, everything created by nature—*amounts to approximately 2.5 trillion tons.* In other words, it took us just a few thousand years (and mostly just the last hundred years) to create twelve times as much as nature created over billions of years—and the rate at which we "create" keeps increasing! Does this not constitute a warning sign of major transformations, even of the impending crisis discussed above?

* * *

By various estimates, natural biodiversity ranges from eight million to one hundred million species, while the diversity of so-called technetic species, that is, of various types of objects created by humans, already exceeds biodiversity by a thousandfold. This is "what we have wrought" over, primarily, the last one hundred to one hundred fifty years. According to some estimates, we will increase the quantity of these species by an order of magnitude every ten years!

* * *

The international organization Global Footprint Network, or GFN, has come up with a well-founded method for measuring what they call "ecological debt." Using this method, every year GFN designates a certain day as Overshoot Day, or Ecological Debt Day—the date when the resources consumed by humanity that year exceeds the amount of resources that the Earth can replenish in a full year. In 1970, this day fell in December, so there was no ecological debt. Later, beginning in the 1980s (we can see the clear correlation with the beginning

---

6   Liz-Rejane Issberner and Philippe Léna, "Anthropocene: the vital challenges of a scientific debate," *The UNESCO Courier* 2 (2018), https://en.unesco.org/courier/2018-2/anthropocene-vital-challenges-scientific-debate.

of globalization), this debt began to appear, and it has continued to grow ever since. In 2019, Overshoot Day fell approximately on July 30![7]

* * *

In 1970, we consumed about 90 percent of the resources that the planet can regenerate in a year. By 2019, this had already risen to 175 percent. The growth rate has doubled and is constantly increasing. Extrapolation using GFN's methods shows that if we continue to "burn through life" at this rate, our ecological debt in the year 2050 will amount to more than four hundred years of resources! Assuming, of course, that we are still alive . . .

5.4. The Danger of Interfering with Human Nature

There are possible conflicts associated with interfering not only with external nature, but also human nature. These stem from technological progress itself. For example, information and communication technologies (ICT) and artificial intelligence technology (AI) open new possibilities for human interaction. Even now, a significant part of human communication has been relocated to the virtual spaces of computer networks. Interaction in these spaces takes place not directly between people, but through their virtual imprints, virtual clones ("avatars," profiles, accounts . . .) which are sometimes radically divergent from their real prototypes.

* * *

Is this a good or a bad thing?

Ethical evaluation ("good or bad," "good or evil") is more than appropriate in this case. At stake are the moral problems of a world in which people can solve creative, informational, and cognitive problems by transferring all routines and second-tier functions to virtual personalities. If such virtual personalities were furnished with AI systems, they could take over the collection, processing, and sorting of information flows, for example. Self-learning artificial intelligence can absorb new knowledge and even apply it to new objects. However, AI is not

---

7   From an interview with Pope Francis: "The fact that has shocked me the most is the Overshoot Day: On July 29th we used up all the regenerative resources of 2019. From July 30 we started to consume more resources than the planet can regenerate in a year. It's very serious. It's a global emergency." Domenico Agasso, Jr., "Pope Francis warns against sovereignism: 'It leads to war,'" *La Stampa*, August 10, 2019, https://www.lastampa.it/vatican-insider/en/2019/08/09/news/pope-francis-warns-against-sovereignism-it-leads-to-war-1.37330049/.

capable of discovering previously unknown knowledge, so there is no need to worry about its competition with humanity on a species level (whatever might be said of its present competition with specific human professions).

But who, how, and for what will this virtual world be used? How will its rules be defined, and to what ends will communication in virtual space be subordinated? Things could reach the level of genuine "virtual horror," as science fiction has energetically explored for years.

* * *

The technosphere has turned into a colossal force which in many ways is already independent from humanity, which further enhances human beings' obligation to bring this force within reasonable limits, forestalling the elementally destructive effects of technogenic processes. If we recognize this obligation, it could spur us to create networks and systems for collective action. But the obligation may go unrecognized, or it may be recognized but remain unfulfilled, due to humankind's collective irresponsibility.

* * *

We thought that we were doing clever things, but in fact, if we continue down our current path, we will have been preparing a change of civilizations. Future civilization might include Morlock-people, à la H.G. Wells, or people with wheels instead of feet—or intellectually advanced people of the noosphere. The technetronic version of the future would have cyborg-people. Or rather, not people at all, not as we currently understand them. Intelligent beings, yes. But not us. Perhaps they would have a different, more "rational" logic of development, and perhaps human beings would not fit into this logic. I may be wrong: after all, this is just a hypothesis. But what if I am right? Facts are stubborn things . . .

* * *

And here, a line must be drawn. Why? Because satisfying needs changes the nature of human beings—their physical essence, for example. What does an athlete do if he wants to lift one hundred fifty kilograms? He trains his body. Perhaps new technology would allow him to do this just by taking pills. This would make it possible for him to take a swing at lifting two hundred kilograms. But why stop at two hundred! Let's alter his DNA to give him an elephant's muscles—easy! Now three hundred kilograms are well within our reach . . . And so on.

* * *

Taking this peculiarity into account, meeting needs without rationally separating real needs from simulative ones may not just radically change isolated qualities of human beings, as biological creatures with both natural capacities and limits to consumption; it may distort human nature itself. This is not science fiction. The technological requirements for events to develop in this way are being created right now.

* * *

For instance, researchers at the Massachusetts Institute of Technology (MIT) in the USA are already altering genes inside human embryos, removing or disabling some, and adding others! Another American institution, The Scripps Research Institute (TSRI), has gone even farther. The four nucleotides that make up natural DNA—adenine, thymine, guanine, and cytosine, which serve as the basis for all life, from bacteria to whales—have been supplemented by researchers at TSRI with two new artificial nucleotides, which do not exist in nature. They have inserted these two interlopers into the DNA of living cells and made them successfully reproduce themselves, passing on acquired (embedded) traits through inheritance, resulting in semisynthetic proteins![8] So we may soon see men with not just elephants' muscles, but "hydraulic power steering."

* * *

But if a human being wants to change his nature as a human being, then are we still referring to a human being as a biological and sociological creature, or are we discussing a different type of being? If we are talking about a human being, we must associate this concept with reasonable limits that would not allow it to develop in such a way.

* * *

The development of a new type of production with unprecedented knowledge-intensiveness, the growing might of technology, and the tremendously broad capacity to satisfy needs are accompanied by the formation of a certain *new type of human*. What will this human be like? The answer to this question is by no means predetermined. Already, there are multiple evident versions of how humanity could develop under the new industrial civilization.

* * *

---

8   Scripps Research Institute, "First stable semisynthetic organism created," ScienceDaily, January 23, 2017, www.sciencedaily.com/releases/2017/01/170123214717.htm. For more details, see Yorke Zhang et al., "A semisynthetic organism engineered for the stable expansion of the genetic alphabet," *PNAS* 114, no. 6 (2017): 1317–1322. [Bodrunov cites Russian sources reporting on these U.S.-published articles.—*Translator's note.*]

The current state of technogenesis is bringing humanity into the extremely perplexing and badly managed world of a technosphere that has evolved according to its own laws. The social order based on capitalist relations of production, which chiefly focuses its production goals on monetary earnings and other indicators of volume and value (such as GDP), is not particularly inclined to settle accounts with the risks and threats that result from subordinating technology to the extraction of profit.

* * *

Will humanity adequately respond to the challenges of the new technetronic or technogenic civilization? Will it come to a new society of humanistic values and widely disseminated "knowledge-creating" activity by people, a society that lives in harmony with nature and overcomes social conflict, in which people's main occupation will be the acquisition of new knowledge? Will this be a society in which material limitations will not play a major role, since the private or personal appropriation of material goods will have lost its paramount importance due to the general accessibility of ways to meet life's material needs? Or will we encounter the opposite?

* * *

People in developed countries, intoxicated by the nearly limitless ability to raise the level of their satisfaction of needs, may be tempted by overconsumption. Less developed countries, due to the former chronic underconsumption of billions of their inhabitants, face the threat of turning their new technological capacity toward unrestrained quantitative growth, producing material goods beyond all reason. Both these tendencies are fraught with the inflation of irrational, fictive, simulative needs.

* * *

A kind of *homo consumericus* is becoming widespread, always chasing fictitious goods with no consideration for anything else. The pressure on Earth's resources will rise, despite the possibility of substantially decreasing resource-dense production. Unrestrained consumerism threatens to swallow up any conceivable quantity of natural resources and suffocate the Earth with waste, while plunging humankind into a vortex of conflicts over material goods and the diminishing resources needed to produce them.

* * *

Emerging now is a world of alienated human beings—alienated from other people, from society, and ultimately from their own essence. Humanity is becoming dehumanized, turning into pseudo-humanity, and threatening its own and its habitat's existence. Alien on Earth. Alien to all. No need to turn to science fiction or await the arrival of aliens from outer space. The aliens are already here. Many people on Earth are already being drawn into the sinkhole of thoughtlessly chasing a fictitious growth in consumption, though the resources devoured by this growth are all too real—not just natural resources, but the body and soul of human beings themselves.

Can we avoid this dead-end path?

5.5. Rising Risks Are Inevitable under the Framework of the Currently Existing Economic Order

Our society has not yet "grown up enough" to make correct use of technological progress and its achievements. Part of the reason why is that technological progress has not yet "fed" everyone. Why is this the case, even though enough grain is produced in today's world to make bread for everyone, while millions and billions of people starve? Because these goods are still subject to the capitalist mode of appropriation. When technological progress is combined with finance capital, which devours its results, all it can do is redistribute income towards finance, not towards productive capital or the satisfaction of people's real needs.

* * *

Finance capital is ready to enrich itself at all costs, including by "ripping off" other individuals, peoples, and nations in the old-fashioned way. At the expense of margins of innovation, at the expense of new markets, at the expense of simulative indoctrination about what to buy, finance tries as hard as it can to establish growing financial profits as the ultimate good. A similar situation emerges at each new stage of technological progress. We are transitioning to a new technological paradigm, but take heed: its corollary is greater expansionism, more wars, more conflict, and so on, even though one might assume that the satisfaction of needs should allow people to live better. Why is this happening? Because we are dealing with a disharmony, the discrepancy between the capacities of technological progress and lagging social consciousness.

* * *

Why is our current situation so tense? What has made the present more fraught than the past? The answer is that each phase of technological progress provides far more new abilities than did the previous one. If these abilities are used incorrectly, then the level of risk dramatically rises. Right now, technological progress offers practically any terrorist the capacity to get an atomic bomb. If society does not mature—precisely *as* society—then, as society, it may create its own direst threats.

* * *

Our social and economic system is extremely interconnected, and it develops dynamically. But as the system's elements develop and interact, they influence one another, while each of them develops at a different rate. Dysfunctional disharmony of these rates, the non-concurrent rates of development of multiple elements of the system, may cause the whole thing to burst, since the strain placed on the links between the elements cannot be unlimited. There is always a limit.

* * *

With every transition, technological changes have led to transformed technological paradigms. Each time, paradigms of production have formed a new type of society: the industrial mode of production and the new technologies of that stage led to the formation, in turn, of capitalist society, rather than vice versa. Now, each new stage offers new capacities, much broader ones, for meeting human needs. If we are saying that we will satisfy human needs at a much higher level in the future than we do right now, and if these needs are not reasonable ones, then—figuratively speaking—we will be using technological progress like an instrument given to a child, or to an underdeveloped creature.

* * *

Humanity now finds itself in this position once more, but the possibilities afforded by the current stage—which are exactly what define its uniqueness—are so gigantic that they may simply lead us straight to the edge of catastrophe if they are improperly used.

* * *

At a certain point, the development of commodity markets led to the birth of finance capital to service them, turning money into the "master," the sovereign, of economic relations. After that, money and financial markets, both due to their basic nature and their unceasing need to expand their sphere of activity and capture new "zones of influence," began to decisively affect the structure and infrastructure of so-called international trade. Once corporations seized control of national markets, they gradually escaped national boundaries. The resulting multi-sectoral conglomerates and transnational business structures, "intertwined" with one another and "sanctified" by the flow of capital, became the foundation of the global marketplace . . .

* * *

It is finance capital that now dictates to political actors how policy may affect the various spheres of social life. This is how we get all kinds of associations (in the word's original commercial and economic meaning), trade wars, pseudo-democratic sanctions, and more.

* * *

The process of globalization is linked to the technological development of society, human society, and civilization. Moreover, in a certain sense, the development of globalization is predetermined, under the "zoo-"paradigm, by human society's technological development. Why? Because, by presenting capital with possibilities for expansion while humanity remains "under-cultured" as a whole (insufficiently "noo-," that is, not yet able to limit its simulative needs and desires), technological progress exists to serve finance capital, which is nearly unlimited. Because technological progress allows capital to be more effectively transferred, used, and so on, while simultaneously fulfilling capitalists' desire to multiply their capital, realizing capital's ability to expand.

* * *

By now, the satisfaction of simulative needs has reached its critical point. Having become a weapon of finance capital, technological progress is creating new but ever more simulative needs, needs which it immediately satisfies, ensnaring everyone and everything in the process. From one perspective, products must be distributed somewhere; from another perspective, the option must be created for people to take these products. But from yet another perspective, this process uses resources—those that are left, that still exist somewhere, that have not yet been set in motion—to drum up demand for these same products as effectively as possible, for the ultimate purpose of growing finance capital.

* * *

Historically speaking, we are approaching a point at which the global process of financialization will have captured basically every possible arena of its existence. It no longer has the option of dramatic territorial gains or extensive growth; nothing remains but the intensification of the pernicious process of financialization itself. What kind of intensification? Mainly in the exploitation of nature, including natural resources, raw materials, etc. Whatever the damage done.

* * *

Financialization will not stop here; it will creep into the human soul. Why? Because, as if in passing, it will create needs that human beings had never previously imagined, which will in fact be simulative needs: in this way, it will bring humanity to ruin. When finance capital makes its way into the social sphere, it changes how people relate to one another, demanding that they consume more, that they absorb themselves in mass culture, and all sorts of other things that are not in fact very important for human beings. This all develops artificially and is hammered into people's heads by global capital.

* * *

The link between technological progress and finance capital, or rather, globalization via finance capital, has certain consequences. It assumes that certain conditions are superimposed onto society's normal existence. Consequently, we see the formation of structures for promoting finance capital, etc. For example, I oppose the creation of new rules for international trade in its current form of existence—precisely because contemporary international trade is a mechanism for promoting the interests of finance capital and imposing new simulative needs on people, even as its highest priority.

* * *

Every modern product is the result of processing tons of natural substances. Producing one pair of shoes requires between ten and thirty tons of fresh water. Just imagine how many pairs of shoes rest unused in stores or are bought and worn only once for a special occasion! Ksenia Sobchak reportedly has an entire room containing nothing but cupboards full of hundreds of shoes. Let us contemplate how many planetary resources, not to mention resources of the human soul, were used in the creation of that room.

* * *

We can take Cambodia as a living lesson. Rubber trees (*Hevea brasiliensis*), introduced into Cambodia's lush jungles a century ago by capitalist colonizers, turned into mighty plantations. Hevea trees produce latex for the rubber industry for twenty to thirty years, and must then be abandoned, at which point the plantation becomes a graveyard of dead trees. This requires the constant felling of nearby jungle every year to plant new trees and grow plantations outward. This is how global finance capital "strips the soul" and "squeezes the life" from whatever territory falls into its clutches!

* * *

The United Nations predicts that by 2030, Cambodia will no longer have any natural forest! Meanwhile, Cambodia is rapidly turning into a land of casinos, banks, shadow capital, sex trafficking, and other such pursuits. The local population is either migrating or growing impoverished in Cambodia's cities, whose development is "soaring" like mirages in the air. (Cambodia's rate of GDP growth has formally been higher than China's for many years!)

This is how finance capital operates: both destroying the rainforest itself, with its irreplaceable beauty, flora, and fauna, and destroying Cambodia's nature, society, and soul. There is a profound connection between the two.

* * *

If we do not supplement our growing technological knowledge with other types of knowledge—about the need for reasonable self-limitation, about the use of "noo-approaches" to organize our lives, and most of all about the new possibilities of technological progress—then we will obviously come to disaster. Ahead lies the point of singularity of our civilizational development. We may not notice when we pass this point, but we will soon feel the consequences. Our choice is between continued "zoo-life," with its "zoo-economy" or "zoonomy"— in which case we will surely meet the fate described above—or emerging from the present into NIS.2 and gradually creating a world of reasonable needs and reasonable production: that is, nooconsumption and nooproduction.

Step Six

Nooindustrial Production

6.1. The Technological Implementation of Knowledge

When we discuss the need to make production rational, we have in mind the application of reason and the knowledge obtained by reason, simultaneously to technological production and to the goals of production, to the needs that production is called upon to meet. Production has always relied on the use of knowledge. What makes modern knowledge-intensive production unique in this regard? Just that knowledge is paramount compared to production's material costs, as we have mentioned already?

* * *

The technological application of knowledge is growing substantially more complex. For some time, a transition has been underway from empirical knowledge to the application of scientific knowledge. "Knowledge acquisition," and application of knowledge, have emerged as their own branch of modern production. At the current stage, interdisciplinary research, and the transition it supports—from the "coexistence" and interaction of various technologies, witnessed earlier, to their convergence, that is, the formation of hybrid technologies with synergistic effect (an effect that exceeds the sum of its parts)—are becoming dominant. These effects lead to the transition of technological and socioeconomic development to a new stage.

* * *

It is the effect of technological synergy that facilitates civilization's accelerated movement along the path of industrial progress. But more important still is the "second-order" phenomenon that arises from this synergistic effect: namely, the development of modern technology and its increasing knowledge-density brings with it an increase in technology's synergistic abilities. In other words, to define a new term, technology's "synergistic density" (or, if you will, its "synergistic power") is growing. This already creates a technical and institutional base for the "accelerating acceleration" (the second derivative, as it were) of scientific and technical progress.

* * *

We now witness this phenomenon as a fully evident practical process. Each new technological solution, in the spirit of using the mechanisms mentioned above by dramatically raising the synergistic density of newly engineered technologies, also improves all the positive parameters of industrial production by orders of magnitude: it lowers the resource-density/costliness/ecological footprint of production, it raises the productivity of labor and the quality of output.

* * *

Labor is the direction of effort, of any human effort, to acquire or apply knowledge that we require to satisfy needs. Even the "application" of knowledge is a form of knowledge! No less than the hands, the head works constantly to meet needs. People now refer to "knowledge production" and the "knowledge economy." I propose that these terms are only acceptable as stopgaps, to clarify the process of "excavating" and using concrete knowledge in today's "economy." *No one produces knowledge; knowledge objectively is, it exists in absolute terms, outside of us.* Our labor is not the "creation" or "production" of knowledge, but the cognition and discovery of concrete bits of knowledge; step by step, more and more, "expansion of consciousness" and of the knowledge-space that is accessible to a person (and to humanity as a whole) takes place; but this is not the "invention," nor in any way the "creation," of new knowledge.

* * *

At present, the process of "knowledge excavation" is subject to the same technological shifts as all other human activity. In recent decades, scientific activity (in all its aspects—from its organization, to its cost, to its results and its implementation within a system of social needs and interests, etc.) has undergone radical changes, so much so that the time is right to discuss its definite transformation: from academic research as traditionally understood, to something new which rises to the challenges of the process of social transformation that is linked to the beginning of civilization's transition to NIS.2.

* * *

The external features of this transformation are obvious. In the twentieth century alone, the number of people directly employed in the scientific sphere grew sixty to seventy times by some estimates. Over the same century, the amount of money spent on science grew by more than one thousand times. Moreover, hardly anyone today will dispute the necessity and inevitability of further increasing this sum.

* * *

As a phenomenon of modern social life, science is more than just a receptacle of knowledge. The development of new industrial (knowledge-intensive) material production in the second generation of new industrial society is based on the prioritization of knowledge in every aspect of creating a new industrial (knowledge-dense) product. As a result, the role of scientific knowledge as a neo-industrial resource grows ever greater, gradually becoming the basic resource of the new generation of industry. This is what determines the observable "external" transformative effects in the development of the scientific sphere.

* * *

This is particularly important to grasp in terms of researching the most important question for the further development of NIS.2: the transformation of the integrated "triangle" of "production-science-education" that constitutes one of the cornerstones of NIS.2's conceptual platform. More and more, the fundamental place in the "triangle" belongs to science, which is becoming the driver of knowledge-intensive production. Factually speaking, knowledge, by becoming production's essential resource (a "direct productive force") is, to a significant degree, supplanting "hardware" in new industrial production.

* * *

The emergence and development of industrial production was closely linked to the transition from individual artisanal labor to mass production. The "capitalization" of industrial production did not just alter social relations; it led to changes in the productive sphere itself. Will the development of science repeat this trajectory, now that it has become the basic resource and dominant driver of new industrial production's development?

* * *

Yes, many facts confirm the following thesis, which seems to grow ever more obvious: to a considerable extent, the scientific sphere is repeating the productive sphere's path of development. We observe the same tendencies, such as the transition from scholars' individual labor to "mass science," the concentration of "scientific powers," to speak in terms of the organization of scientific work. From the economic point of view, we witness the "monetization" of science and the "capitalization" of the scientific sphere. The results of scientific research are transitioning from pure products of the mind to scientific commodities, susceptible to all the usual methods of market turnover. The creative act of obtaining individual scientific knowledge is becoming the "production of scientific output," whose nature is increasingly utilitarian and goal-oriented.

* * *

At the present stage of social production's development, this process is objective and inevitable. So long as productive forces operate within the boundaries of currently existing social relations, science, becoming a "direct productive force" and taking on the role of a foundational resource and, therefore, foundational capital, will inescapably repeat the developmental trajectory of any resource that constitutes the basis of capitalist relations. For this reason, we can foretell many future twists and turns in this path and anticipate many problems.

* * *

Many problems, but not all. The concept of NIS.2—which assumes not only the development of new industries as a means of material production on a qualitatively new technological basis, but also the qualitative transformation of society's institutions—takes as its starting point the "special nature" of knowledge as a social phenomenon. This "special nature" amounts to the fact that, unlike material resources, knowledge, whatever imaginary ownership claims and personal boundaries might be imposed on it, is in principle "reproducible," non-private, non-individualized, etc.!

* * *

The progression and development of knowledge acquisition in a more socialized direction will continue. Moreover, we can observe a tendency toward the increasing significance of knowledge within the labor function, giving possessors of this "sacred" element the upper hand, which in principle allows for their liberation out from under "the power of capital." We can already observe the repeated manifestation of this incipient tendency: alongside the laborer's classical dependence upon capital, the reverse situation—the employer's dependence upon a worker endowed with rare and important abilities (that is, knowledge, embodied in the labor function)—often arises. Moreover, this "reverse" dependency is often much stronger than the "original" dependency.

* * *

The colossally growing role of knowledge, as embodied in the technologies and products of production, is bound to make itself known in highly significant fashion—not only in the growing possibilities for satisfying human needs, but even in the process of shaping those needs. Under the systemic conditions of the modern economy, this influence is contradictory. Even with the rise in possibilities for meeting ever-growing needs, which seems purely positive, things are not as simple as they appear to be.

6.2. The Double Role of Knowledge in Shaping and Meeting Needs

What's wrong with the unlimited growth of consumption? After all, with such expanded capacity to satisfy needs, even those needs that appear simulative, false, or illusory at the current level of productive development may change categories and become non-simulative. However, the growth of simulative needs still needs to be limited, lest it lead to meaningless waste of resources and distort the further formation of needs.

* * *

It will not do to brush off this problem merely by gesturing at Maslow's pyramid, claiming that so long as vital, material needs are satiated, then our knowledge of more elevated needs will increase "all by itself," since the needs we have already met will no longer motivate us. Maslow's pyramid explains nothing; it merely codifies a few empirically observable tendencies.[1] The

---

1   See A.H. Maslow, "A Theory of Human Motivation," *Psychological Review* 50 (1943): 370–396. Maslow's conception has been seriously criticized. Many believe that the interaction of needs

reason why such a major shift occurs within the structure of needs, and the question of what problems emerge over the course of this shift, must be dealt with separately.

* * *

Maslow's main error (shared by some of his followers and critics) is his attempt to explain changing needs solely *from within* the individual human psyche. In reality, this is not just an "individual story." The patterns that appear when certain needs are replaced by others can only be understood with reference to collective social phenomena, and the causes of these patterns must be found in the basic factors that determine people's lives.

* * *

Why is it and when is it that once society enables the satisfaction of demand for the basic needs of life on a mass scale, these basic needs are displaced by others? This is less a factor of the degree to which basic needs are sated (satiety is a *condition*, but not a *cause*; it is only once vital needs are sated that the transition may occur, *but it also may not occur*) than of the changing character of people's fundamental labor activity.

* * *

Within labor activity, the creative function is growing—rather unevenly, but proportionally to the progress of knowledge-intensive technology that makes it possible to meet the needs of sustaining life more fully. Much more than the sphere of consumption, it is production that dictates the need for people to be a) creative and b) responsible (by virtue of the potential power of the technosphere that they set in motion).

* * *

For this reason, the shaping of "cultured people" is becoming more and more important for our economic future, determining the growth of intellectual and spiritual requirements. This emphasis is very important as we move toward NIS.2.

* * *

---

at different levels is far more complex than Maslow assumed (which is to say that one can move either up or down the pyramid). See, for example, the theory of a hierarchy of needs advanced in Clayton Paul Alderfer, "An empirical test of a new theory of human needs," *Organizational Behavior and Human Performance* 4, no. 2 (1969): 142–75.

The combination of knowledge and cultural imperatives is essential. After all, a human being's reception of knowledge about the concrete means for satisfying a concrete need—in other words, the means for resolving an emergent conflict between needs and possibilities—is not limited to a single vector. Knowledge, by virtue of its universality and endlessness, does not provide a single, unique means for satisfying such-and-such a need in such-and-such a way, but rather an endless multiplicity of such means. The task of an individual person is to choose the most optimal/appropriate option; for this purpose, man is endowed with the power of knowledge, and also with free will, the ability to decide which option to choose.

* * *

That said, man determines the optimal criteria for choosing a concrete path to resolve a concrete conflict based on the extent of knowledge that is available to him within the specific subsystem of his existence.

* * *

The necessity of satisfying humanity's escalating needs is what drives development. Yet man cannot meet his needs without resorting to the technological application of knowledge. In early periods of civilization's development, humanity primarily relied on knowledge as a product of empirical experience, but by now, one cannot do without the large-scale "production/extraction" and application of scientific knowledge. Moreover, human needs can themselves only be cognized or clearly formulated with reference to knowledge obtained by human beings. At the same time, new knowledge makes it possible to discover, articulate, and satisfy new needs.

* * *

It is the application of knowledge that distinguishes human labor from the instinctual activity of animals, and *precisely because of knowledge, material production constitutes human beings as social beings.* Man is what he does; man is what his activity is.

* * *

The fact that man's productive activity determines his character as a social being was emphasized by Karl Marx:

> "As individuals express their life, so they are. What they are, therefore, coincides with their production, both with *what* they

produce and with *how* they produce. The nature of individuals thus depends on the material conditions determining their production."[2] ". . . The final result of the process of social production always appears as the society itself, that is, the human being itself in its social relations."[3]

Further positions regarding the dependence of humanity's social being on its activity were developed by Soviet philosophers and psychologists.[4]

* * *

Yet material production is an activity based on knowledge. Man can only produce if he has knowledge: when he enters the production process, he necessarily acquires knowledge, and by the time he leaves this process, he is enriched with new knowledge. It is obvious that this fact plays an enormous structural role.

* * *

Only human beings possess the unique ability to discover and formulate traits, laws, and patterns of appearance in the external objective world. And only human beings are capable of applying the knowledge they have discovered to transform the external world, using this knowledge to define suitable objects (materials), create means of transforming them (technologies), and formulate for themselves the goals of such transformation—the creation of products that meet human needs (broadly speaking, including both material and immaterial needs). Without knowledge, it is impossible not just to create something new that does not exist in nature, but also even to make a copy of something, since having the idea to make a copy is already a creative act.

* * *

In coming to know the world, man also comes to know himself as part of the world. The development of man's needs will inevitably lead to the development of the need for man to know himself, his place in the world, his interests, and the "interests" of the elements of the world surrounding him—namely, inorganic

---

2 Karl Marx and Friedrich Engels, *The German Ideology*, ed. C.J. Arthur (London: Lawrence & Wishart, 1974) 42.
3 Marx, *Grundrisse*, trans. Nicolaus, 712.
4 See G.S. Batishchev, "The Active Essence of Man as a Philosophical Principle," in *The Problem of Man in Modern Philosophy* (Moscow: Nauka, 1969): 73–144; A.N. Leontiev, *Activity, Consciousness and Personality* (Moscow: Politizdat, 1975).

and organic nature, both of which are related to him. Finally, he starts to account for these "interests" in the process of satisfying his own needs and resolving conflicts between himself and the world. Man comes to set himself apart from the surrounding world and conceive of himself as a separate being. This is the process through which individuality and social communality are formed.

* * *

By making one decision or another in the process of resolving conflicts, man becomes an individual and perceives other members of society also as individuals with their own interests, right to choose, and corresponding abilities. The solution of many intersecting problems leads to the necessity of meeting a multitude of needs (some of them contradictory!) concerning the relations between different members of society. Social relations emerge with the development of the social whole, along with the development and cognition of the so-called "public interest."

* * *

Thus, social relations are conflicts that are resolved within frameworks—rules, norms, laws, etc.—established by society itself (based on social "consensus" or agreement, which in turn is based on society's comprehension of the optimal conditions for the social collective, and on its experience of knowledge accumulation!) Violating these rules (straying outside the bounds of the public interest) is perceived by society as a destructive act that destabilizes the system of social organization and corrodes its structure.

* * *

Consequently, the need arises to create natural constraints against humanity's ability to make highly suboptimal decisions. Once again, these constraints are shaped and reinforced throughout the fabric of human culture based on the acquisition of specific knowledge. At the same time, man can develop and apply technological solutions that enable the creation of favorable conditions for making optimal choices.

6.3. Technologies of Trust

Any civilization or social order will give rise to a particular system of relations, principles, morals, traditions, rules, habits, and so on. We, who have generated many "agreed-upon" elements within our own culture, must necessarily place our trust in them; otherwise, there would be no structure or home that we could call "our own civilization" or "our own cultural space."

* * *

The violation of one of these elements by a member of society is perceived as going beyond the boundaries of such a shared cultural space. Violations on a mass scale lead to the destruction and transfiguration of this space. For instance, we describe the deception of one person by another as the violation of a cultural tradition, namely, of our trust in one another. Because we constantly rely on the benefits and incentives of civilization (commodities, services, relationships), we are obligated to regularly check them (typically by using technology) against certain stated criteria of an accepted civilizational code. More often than not, though, we must simply rely on trust, due to the impossibility of subjecting everything to thorough examination, reconciliation, alignment . . .

* * *

Trust is one of the basic problems of contemporary civilization. At the very least, its economic significance is exemplified by bank transactions: the costs of establishing compliance for transactions amounts to roughly half of all expenditures within the banking system. The trust factor only grows in importance for preserving both civilization's basic principles and its stable development, since technology is more and more capable of "wriggling into" any one of us. The ability to violate social conventions grows ever higher, while our defenses grow ever weaker!

* * *

Thus, the growth of society's technological "armament" is accompanied by the problem of raising the general level of trust in social relations (let he who will say that trust is *not* one of the most important elements of cultured community cast the first stone at me for saying so!). To resolve this problem, we should not prioritize acculturation as such (acculturation, though necessary, plays a secondary role) but rather the transformation of the conditions under which social relations are realized. In other words, we should prioritize technological change.

* * *

For just this reason, "technologies of trust" should be one of the key components of the technological basis of the future: linked with both the core criteria of economic rationality and the zoological side of human nature, these technologies will allow us to make decisions untainted by egotism.

* * *

It is imperative that we set the creation of technological means for raising levels of trust as a priority of technological development. For instance, if deception cannot physically take place (especially given widely accessible means for the simple satisfaction of needs) then attempts at deception will come to naught! If deceit becomes technologically impossible, then why not believe the information you receive? The universal implementation of "technologies of trust" will gradually change our cultural codes: our habits, concepts, ways of communicating, and more.

* * *

Let us note that "technologies of trust" have existed and have constantly been developed all throughout history. Nowadays their development is rather intensive, due to the growth of social demand for them. The entire world is gradually starting to use distributed database technology—a.k.a. the blockchain—which underpins virtual currency (cryptocurrency). This is because such technology raises the overall level of social trust.[5]

* * *

Trust is the key word here. If we choose a path that increases levels of trust— such as by using technological procedures that guarantee the authenticity of documents with ironclad certainty—then we will no longer need to waste time establishing trust. In other words, we will be more able to spend time on other activities.

* * *

For example, if we have already grasped, understood, and accepted the truth of an initial group of theorems when we begin to solve a mathematical problem, we may then construct other theorems without going back to the first ones and proving them all over again; we can rely on what we consider to be true. If we construct our relations with one another on the "impossibility of deceit"—taken as an element of human upbringing—then after two or three generations, humanity will forget the meaning of the word "deceive": we will

---

5 See Don Tapscott and Alex Tapscott, *Blockchain Revolution: How the Technology behind Bitcoin and Other Cryptocurrencies Is Changing the World* (New York: Portfolio, 2018).

no longer understand *how* to deceive, and the necessity of deception will wither away. (See fig. 6.)

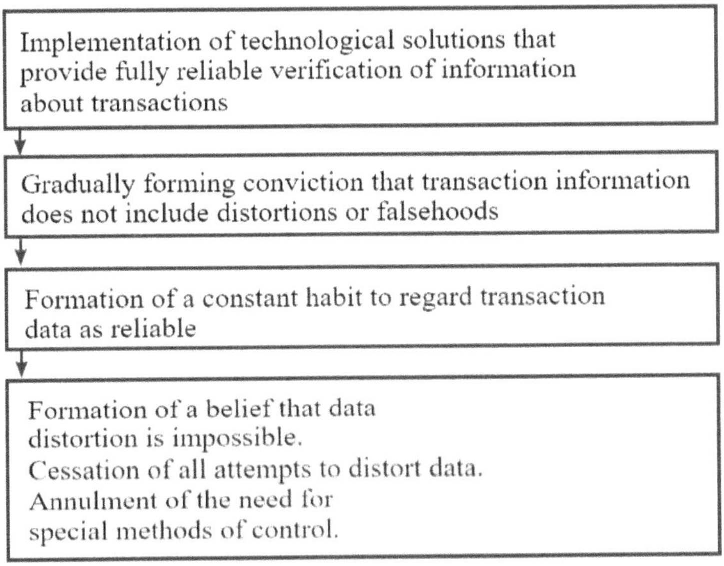

Implementation of technological solutions that provide fully reliable verification of information about transactions

↓

Gradually forming conviction that transaction information does not include distortions or falsehoods

↓

Formation of a constant habit to regard transaction data as reliable

↓

Formation of a belief that data distortion is impossible.
Cessation of all attempts to distort data.
Annulment of the need for special methods of control.

FIGURE 6. The cultural and educational significance of technologies of trust

* * *

Each person's *level of trust* is exactly what determines their individual threshold for believing in the truthfulness or untruthfulness of particular phenomena. To a considerable extent, this shapes our knowledge space as a "*trust space.*" For this reason, we say that we need to increase the level of trust right now, because *the greater the knowledge sphere/space, the greater the level of trust.*

* * *

At one point, Francis Fukuyama wrote about the "radius of trust" that surrounds people. My initial radius of trust is that I trust my neighbor first and my family second (or the other way around)—and I trust both more than I trust my municipality (and much more than I trust the state) and so on. (After all, placing trust in an abstract state is not an easy thing to do!) However, Fukuyama neglected to say the most important thing: within this space or "radius" of trust, people do or do not place their trust in the correctness of certain criteria and norms.

* * *

The radius of trust depends precisely on the overall level of trust. Do 80 percent of people believe something? Or 30 percent? Or does everyone categorically not believe in it? If people do not believe, they will strive for unvarnished truth and absolute knowledge. Even to the point of dismantling the family, or the state itself, if people have lost their trust in them. These levels of trust can be expanded and strengthened, including through the implementation of new technology. Such technology will make it possible to gauge each issue more effectively. Yes, or no? True, or false? To believe, or not to believe? By answering these questions, we can decide on a correct and reasonable order for any given portion of space, technological solution, or social phenomenon.

* * *

In this sense, the blockchain (one of the many "technologies of trust" created by humanity, and the most prominent one in the present moment) may do much more for the development of democracy—and for more rational social development, advancing toward noosociety—than dozens of other methods for raising our faith in the state. This is not only (nor even primarily!) because the blockchain "verifies" the results of choices, but because it may give us the ability to elect the leaders that are most qualified to rule (according to the basic criteria of the moment).

* * *

Technology is merely the emergence of knowledge into the real, material world. We all know that $2 \times 2 = 4$, and it no longer matters how we learnt it. We have verified it many times, and by now we believe that it is the truth. No one could tell us that $2 \times 2 = 5$. We would not believe them. In just the same way, we place much more faith in American elections than in other countries' elections; we know that American banks are technologically more well-protected than other banks; Americans have more legal protections when they make deposits, and they are harder to defraud, because "fraudsters" are highly likely to be punished by law. In addition, American trustworthiness is better secured by technology. For these reasons, we all have more faith in their currency and in their elections.

* * *

After knowledge of what is and is not true comes certainty about which of our notions are truthful. This certainty is reinforced by frequent repetition in practice; as the saying goes, the result of an experiment that has been confirmed many times must be recognized as the truth. And though this truth may be overturned by new experiments in the future, at the present stage our heuristics and our knowledge permit us to regard these assertions as truthful.

* * *

Technologies based on verifiable laws of the physical and material world surround us on all sides, and we ascertain and trust in the rectitude of criteria and parameters, of concepts formed in our heads, that rest on this basis. Thus, developing these technologies today, we may come to the point that humanity increasingly understands the world, including the part of the world that allows human beings to give up their simulacra. Recall the humanists who spoke of human values, eternal truth, and so on, introducing these ideas into the basic criteria of our knowledge and of our culture. Nor did these things appear in a vacuum: they were based on apprehension, interpretation, cognition, acknowledgment, and knowledge of what is incorrect and what is correct. A transformation occurred in basic criteria of what is good and what is bad.

* * *

Thus, the development of social relations of production on the basis of advanced technological methods for verifying economic activity will make economic activity more open and transparent. This enables a dramatic decrease in transaction expenses, which, in the current economic system, are an obstacle to fully unleashing the sharing economy, crowdsourcing, and other types of collective activity and shared use.

* * *

A high degree of openness and a high degree of trust, consolidated within the norms and traditions of human behavior, with a firm technological basis—all this is a vital prerequisite for the evolution of property relations. In the end, the need for private control over resources, their private ownership, will conclusively disappear. Power and domination will evaporate from property relations.

* * *

Not only regimes of shared use will develop, but also regimes of free access. However, this does not mean that the need for institutions that regulate this free access will wither away. On the contrary, carefully calibrated institutional frameworks will be needed to prevent the diffusion and squandering of common resources, even in the absence of private appropriation and its selfish motives.

6.4. The Development of Knowledge as a Cultural Phenomenon

Humanity's social nature unveils an exit from its ripening civilizational crisis. While the external and natural resources of humanity's vital activity—those which are required for its existence as a biological creature—are objectively limited, and require humans to behave quite cautiously, *social* resources present a different situation. The most important resource of human vital activity—the human capacity for knowledge, and the ability to convert extracted knowledge into technology—is limitless.

* * *

These capacities can only be realized with the *simultaneous* transformation of the technosphere that humanity has created and the transformation of our social order in response to the demands of this new technosphere. These transformations will entail geo-economic shifts: the balance of forces in the global economy will not escape change. That said, leadership in the new economy will not solely be determined by leadership in developing and applying the latest technologies.

* * *

The global economy's leaders will be those countries (or their corporate groups, unions, and conglomerates) that not only show their ability to master new types of knowledge, implement that knowledge in technology, and rebuild production based on those technological achievements: a necessary condition will be *an all-new developmental paradigm, a.k.a. the transformation of the goals and motives of human behavior*. In fact, without this paradigm shift, even the current technological revolution will either prove to be impossible or threaten humanity's self-destruction.

* * *

A change in global economic leadership is an all-but-inevitable consequence of technological shifts on a world-systemic scale. And since this transition will be conditioned by our present-day economic systems and models, it is difficult (maybe impossible!) to hope that the forthcoming changes will unfold without conflict. We can predict with certainty that there will be struggles over leadership—so finding a way to soften these conflicts, preventing them from taking on sharp and destructive forms, will remain an urgent task.

* * *

Who can show us the proper path forward? What should be the "design" of such a path? Where is the person who will genuinely start to work through these questions and draw the contours of our path? The whole world is our blackboard! And this task is a universal human task. We should all work at it together; we should appeal to the international community to collectively occupy ourselves with the new industrial development of society. If this happens quickly and consensually, then notorious socio-economic tensions will resolve themselves with as little conflict as possible. This is not an idea meant for any single country in isolation; rather, it is the objective path of our common civilizational development.

* * *

Scholars, politicians, and businessmen have already offered more than enough predictions about the civilizational shifts that await us. Nonetheless, the economic community still lacks a clear conception of the nature of these shifts. Most economists are not looking that far into the future. Others, "soothsayers," who can sense the ground of the economy shifting beneath their feet, rush to calm themselves and others by inventing comforting terms such as "the new normal." Everything is topsy-turvy, economic growth is slowing down, technical progress and productivity are decelerating, the usual levers of control have no effect—but we are meant to believe that nothing much is going on, and that this is just "the new normal!"

* * *

A coherent model of the oncoming future has yet to take shape in the analyses of experts. Perhaps this is caused not by the "limited intellectual abilities" of the community of experts at large, and of economists in particular, but instead by the fact that they do not want to draw awkward conclusions and break with the reality to which they are accustomed? Yet one way or another, this break must occur. The civilizational crisis that threatens us can only be overcome by the power of critical reason that is fearless in its willingness to look squarely at the future's dangers and abandon outdated approaches that stultify our ability to cognize a genuinely new reality.

* * *

A technological leap into the future will only provide humanity with a real step forward if it is based on new principles, noo-approaches, that are alone capable of showing us the right ways to use our escalating (and therefore potentially dangerous, but simultaneously offering considerable new gains) technological potential.

* * *

The noo-approach offers *the unity of technological might with the power of knowledge and with human reason as represented by the traditions of human culture.* Henceforth, it will be society's cultural codes that serve as a vital condition of the technological use of knowledge, *and whatever humankind makes of itself will depend on our cultural norms.*

* * *

In creating a foundation for humanity's exit from direct production, new technological forces form a basis for the withering away of economic relations (that is, the struggle to use and appropriate the resources and results of production). Consequently, though, society itself will undergo profound transformations. The response to the challenges posed by the extensively "technocratic" scenario of development, which leads to a dead end of civilizational crisis, should be conscious intensification of the creation and use of technologies that support both man's personal development and the improvement of the cultural codes of modern civilization.

* * *

The ubiquitous and "stacked" application of these technologies will mean that social institutions change as well. For example, real direct democracy is becoming possible: not just elections, but the direct resolution of all sorts of questions of collective life based on trust-based consensus (requiring no verification!)—whether to allow trams on a particular street, whether to take down a monument, whether to build a factory next to a residential district.

* * *

It is important to emphasize that all technological development, in this version of the future, will be targeted at achieving the goals of rational social development and the satisfaction of the human personality's rational (non-simulative) needs and desires (noo-needs). All production of socially necessary products will be oriented towards meeting rational needs according to codes of civility and culture that are shaped by the framework of NIS.2.

* * *

It does not matter who will do the work: either a robot (most likely) or a human creator (standing "above production"). Material production will remain society's basis, and the means of producing goods will be industrial, based on up-to-date technology. More precisely, the means of production will be nooindustrial, to meet the needs of nooindustrial society existing in the noosphere.

* * *

At first glance, these theses seem largely "Vernadskian." Nonetheless, a deeper reading suggests a different interpretation. Many past thinkers (Karl Marx, Vladimir Vernadsky, Erich Fromm, the theorists of the Club of Rome . . .) have appealed to human reason as the way to solve mounting problems. However, they had no ready-to-hand answer regarding what concrete material means could be used to enact this solution. It seems to us that we can finally provide this answer. We must move from a purely humanistic treatment of ideas related to the noosphere, as can be found in debates over social philosophy, toward grasping the fact that these ideas can become reality—on the solid foundation of the developmental tendencies of material production.

* * *

In this sense, the justification for concepts like NIS.2 and noonomy turns out also to be an approach to justifying the development of the next stage of human civilization. We would propose calling this new stage *noo-civilization*, in which production would be subject to human reason rather than just technology and the technical realm. (However, human reason would itself rest upon the strictly material processes of nooindustrial production, since without this connection, reason would be unable to develop, or even to secure its mere mundane existence!)

* * *

On the same note, the social role of knowledge is rising dramatically: as a means for discovering newer, more effective, and more economical ways to satisfy rational human needs (as opposed to the current path, that is, a purely quantitative escalation of consumption, whose limits are already evident); and as a means for resolving the conflicts and tensions that accompany profound technological and societal shifts. That said, the new society will not be created by technology in itself; the main role will be played by human beings endowed with knowledge, truly rational humans. (This is why we object on principle to technological determinism.)

* * *

It is precisely culture (morality, what people call core values, etc.) that will shape the most important element of this new society's code of civility—namely, human self-control, which will guide people away from the unrestrained quantitative escalation of consumption, weighed down by the pursuit of different kinds of mirages and simulacra, and towards the formation of rational human needs and desires (noo-needs) that give primary place to the *quality* of needs and desired goods. Culture is what will ground the qualitative newness of interpersonal interactions—within both social life and the process of creative labor. At the same time, technological progress turns out to contain enormous potential for the transformation of the cultural codes of human civilization.

* * *

Which social mechanisms will make it possible for us to set goals for production and technical development that facilitate the development of humanity itself, and to make the processes of technological advancement answerable to this task? This is the central question of how social structures will evolve as we transition to noosociety.

* * *

The development of NIS.2 toward noo-civilization is tantamount to the transformation of the standard role of today's commonly accepted basic social institutions—that is, the state, money, and the means of both appropriating social wealth and then gradually frittering it away. The social order will gain stability by basing itself not just on trust, but on firm knowledge that information obtained by "public" exchange is always true and authentic. As we know, there are many different types of "knowledge" out there. But demand will grow for *true* knowledge, *verified* knowledge, *trustworthy* knowledge. Rational knowledge.

* * *

The evolution of human reason's social role is spasmodic. Everything will be determined by what becomes of human reason. Will reason appeal to people's collaboration to reach lofty goals, or will free rein be given to the dark side of the forces borne by knowledge? The education of rational human beings (which is to say, cultured human beings) is becoming a core imperative in the process of forming the society of the future—as is answering the question of how people will manage to work together to reach their common goals.

Step Seven

Culture as an Economic Imperative

7.1. Nooproduction and Needs: The Character of Production and the Character of Needs

Human beings' versatility, formed in response to the challenges of the technological revolution; the creation of new needs and new ways to meet them, also associated with new technologies—where is all this leading humankind? And where is it leading today's economy?

* * *

As shown above, the ability to increasingly satisfy needs at lower and lower cost also creates *both* the possibility of reducing our ecological footprint *and* the temptation towards superabundance. Which one will humanity choose? How will we define ourselves?

* * *

What will it be: will other motives replace personal gain, for example, self-development, quality of communication, or public recognition? Or will humanity drown itself in a sea of ever more illusory and finicky needs and wants? Or will it hide from life in virtual reality and cyberspace? After all, the digital world may expand communicative possibilities, or it may shrink them in the service of human self-isolation. Recall Japan's "hikikomori," people who go for years without stepping away from their computers, rejecting not just social interaction but even the normal routines of life, from regularly eating and changing their clothes to maintaining their physical fitness.

* * *

If we make it past this crossroads and emerge into "nooproduction," then for the most part the latter will constitute *the production of humanity itself, rather than the production of the material conditions of its existence.* Correspondingly, the structure of human needs will also change. The prevailing needs will be those related to self-development, to the spiritual plane, to communication, to public acknowledgement. And these are the needs that will regulate the technologies used in production, the products produced, and the organization of production, for the sake of meeting material human needs. These shifts in the structure of needs will be determined by the progress of human culture, meaning a specific type of knowledge.

* * *

Thus, *social production in noosociety*, insofar as we can judge by analyzing objective processes that began in the recent past and are already underway at present, will take the form of a system that includes:

- the prioritized development of knowledge-intensive "smart" production (which we may call *nooproduction*, eliminating the quotation marks);
- a corresponding integration of production, science, and education within a unified reproductive framework, leading to the formation of a new type of socioeconomic reproduction—*nooreproduction*—that prioritizes the creation of the conditions for noosociety's development;
- the gradually decreased role of utilitarian and simulative needs and the ascent of a new class of needs: the needs of "rational man," or *noo-needs*;
- the concomitant development of new values and behavioral motives, which lose their economic properties, for the main human subjects of material and spiritual production;
- during the transition to this period, economic relations and institutions will transform toward greater socialization and humanization, specifically in terms of the active development of *noo-oriented* economic programming, active industrial policies to prioritize the development of "smart" production, and strengthened public-private partnership to resolve these issues;

- and, last but not least: the ascent of culture as a sphere that solves key problems of *noodevelopment*.

Nooproduction is knowledge-intensive production that minimizes direct human involvement. It is oriented towards the satisfaction of *noo-needs*, prioritizing the creation of developmental conditions for humanity's rise in the sphere of knowledge and culture.

* * *

How can we summarize the objectives of nooproduction if economic goals will have faded into the past? We may define these objectives as **the ascent of the human personality.**

* * *

The ascent of the human personality is the direct goal of production in noosociety, consisting of the development of human qualities and expansion of human beings' cultural environment, as regulated by culturally determined values.

* * *

This ascent will occur through the rising importance of spiritual needs in every area of human culture. An important factor will be the need for conscious self-discipline and the limitation of simulative needs (which, incidentally, when coupled with the new possibilities afforded by technology, will seriously contribute to the resource-efficiency of this developmental path).

* * *

Such self-limitation does not constitute some externally imposed imperative. Of course, during the stage of transitioning to noosociety, a certain role will have to be played by external moral imperatives, explanations, persuasion, and, finally, the nurturing of habitual rational self-discipline. No doubt this will be facilitated by accelerating technological progress, which—via the new industrial production of NIS.2—will cause the progressive devaluation of physical, material, tangible products by simplifying the satisfaction of humanity's vital (and other non-simulative) needs and making material goods less and less important to individual human experience. Instead, meeting human beings' rising spiritual needs will become more and more valuable.

* * *

But of course, *internal* self-discipline, emerging from how the structure of needs is determined by the new nature of human activity and by the social relations that foster such activity, will be the most impactful factor. Even today, for example, people who occupy themselves with decoding the human genome or developing technology to send expeditions to Mars are hardly concerned with acquiring gigantic yachts or villas on the Côte d'Azur, regardless of their level of income. Those who are excited by this type of work find such "needs" irrelevant; rather than helping them, fulfilling these needs would interfere with the pursuit of their genuine goals.

* * *

The thesis offered above, concerning the ascent of the human personality, resonates to some extent with Karl Marx's famous idea that in future society, "the free development of each is the condition for the free development of all,"[1] along with the corresponding position of Vladimir Lenin.[2] In the USSR, these views were regarded as a "fundamental law of the communist system."

* * *

That said, Marxists see free and full development as a law of communism, the result of a successful violent revolution; what we have in mind is *evolution*, and the features of noosociety that we have objectively derived are quite far from the communist schema proposed by Marxists. Furthermore, "development" for us cannot take any arbitrary direction it likes (see the discussion of self-discipline above). Finally, our starting point assumes that different dimensions of personal development are of unequal significance, and we take *spiritual* development as paramount.

* * *

The quality of the spiritual and cultural component of human development is exactly what must determine all the other dimensions of development, subordinating them to the best norms of human culture. However, the progress of human personality is not an end in itself for noocivilization. Noocivilization must develop steadily and sustainably. The system should be sustainable,

---

1   Karl Marx and Friedrich Engels, "Manifesto of the Communist Party," in *The Marx-Engels Reader*, 2nd ed., ed. Robert C. Tucker (New York: Norton, 1978): 491. See also "Critique of the Gotha Program," *The Marx-Engels Reader*, 531; and *Capital*, vol. 1, trans. Fowkes, 613–614, 739.
2   V.I. Lenin, "Notes on Plekhanov's Second Draft Programme," in *Collected Works*, vol. 6, ed. and trans. Clemens Dutt and Julius Katzer (Moscow: Progress Publishers, 1961) 52.

working towards greater sustainability and systemic self-preservation rather than towards rupture.

* * *

Preservation of ourselves as human beings, that is, preserving a system whose basic sustainability rests upon developed, new, important human beings as its foundational element—this is our goal. This is precisely why we need "new people." But any slogan about creating "new people" for its own sake, as an end in itself, would be meaningless. (We recall the Party's dogmas about "scientific communism" during the Soviet era!)

* * *

For this reason, the new person is not a goal in itself. Our actual goal is clear and comprehensible: the "new" person will be an element of noosociety as a system, allowing civilization to preserve and maintain its steady development (which is, after all, the basic principle of all existence). If, instead of "new" people, noosociety were to be populated by "old" people who fit the "old" system, then society would systematically change: it would become technetronic, with the consequences described above.

* * *

As the human personality develops, social relations will only be preserved so long as man consciously understands what he may and may not do, so to speak. We need institutions that do not uphold what happens today, within the framework of global capitalism, but that orient themselves toward securing a noo-version of development.

* * *

For this purpose, however strange it may seem—yet again!—technological development is necessary. In fact, emphasis should be placed not on NBIC-technologies as a whole, but most prominently on C-technologies: nanotech, biotech, and IT were the cutting edge from the 1950s to the 1990s, making them somewhat "worn out" these days. In the near future, we should transition from information technologies to cognitive technologies; otherwise, if we do not pursue the development of human abilities, man's capacity to more deeply know himself and the world, we will fail to absorb an enormous amount of knowledge and will be unable to successfully "implement" the noo-version of development. Only pursuing technological development will allow us to be more or less confident in the future.

7.2. Cultured Man as a Product of Nooproduction: The Ascent of Personality

When labor productivity increases (as needs are reduced or limited), the duration and significance of time spent working decreases, while that of leisure time rises. NIS.2 is already equipped to substantially increase leisure time, but this will not immediately bring about "augmented happiness": first, people must learn to use their leisure time for self-development (elevating their spiritual needs, culture, etc.)

* * *

We can understand the skepticism of Hannah Arendt, who doubted that the expansion of leisure time would lead to greater human development—since, she believed, the reality is that people are inclined to use their free time for nothing but thoughtless consumption. "[T]he spare time of the *animal laborans*," she wrote, "is never spent in anything but consumption, and the more time left to him, the greedier and more craving his appetites. That these appetites become more sophisticated, so that consumption is no longer restricted to the necessities but, on the contrary, mainly concentrates on the superfluities of life, does not change the character of this society, but harbors the grave danger that eventually no object of the world will be safe from consumption and annihilation through consumption."[3]

* * *

Indeed, in the social system in which we currently live, namely, so-called capitalism, this is how things are—because people are granted leisure time only in order to consume what they have produced during their working hours. With equal force, capitalism compels people both to consumption and to production for the sake of that consumption.

* * *

Society may find a way out of this vicious cycle, but not by means of the ideology of asceticism, a.k.a. compulsory rationing and reduction of consumption; nor by propagating higher ideals through mere words; rather, the way out is to reduce time spent on necessary work (which assumes, as a

---

3 Hannah Arendt, *The Human Condition*, 2nd ed. (Chicago: University of Chicago Press, 1958), 133.

prerequisite, the modern industrial mode of production) and to use leisure time to develop creative faculties.

* * *

But this does not mean that Arendt's doubts appeared in a vacuum. *The transition from leisure-as-consumption to spare time as a space for developing human culture will be neither quick nor simple.* This is a colossally important problem which may lead to quite serious, large-scale, and deep difficulties. Only by solving it will we definitively arrive at the epoch of the noosphere.

* * *

In the second generation of new industrial society, man has the considerable capacity to act not as a thoughtless consumer, but as a creative human being. This is true to the extent that the use of spare time for creative purposes depends on the coalescence of the material prerequisites for creative activity: that is, access to means of self-education, means of physical self-improvement, means of scientific and artistic creation, and so on.

* * *

Of course, another necessity is the transformation of the relationship between work time and leisure time, in favor of the latter. And the transition to the next stage, nooproduction, presents humanity (like never before!) with far-reaching and profound issues, concerning the acquisition of new knowledge to ensure a leap forward in technological progress and to grasp the trajectory and limits of our own development. It is precisely the need to resolve these issues, along with humanity's practical engagement with the technological (and socio-practical) application of science, that will determine the face of leisure under noosociety.

* * *

Though Arendt drew her conclusions based on having observed the real social contradictions of her moment, there is one rule that she did not consider: changing the character of human activity in order to gain new knowledge will also—gradually, not all at once—alter the structure and qualitative content of humankind's needs and desires, ultimately *changing how leisure time is spent.*

* * *

It is discourse, information, and knowledge that will become more valuable than even the most precious material values of the previous era. We are not far from the point at which this prospect will be widely understood. The world is approaching the end of a cycle, a turn of the "big wheel," as the ancient Maya Indians used to say. Knowledge, and the words that express knowledge, are now taking on primary importance. Remember the Bible: "In the beginning was the Word." And also, as it turns out, at the end. Of course, every end signifies a beginning. But the beginning of what?

* * *

No doubt the new society will bring fundamental lifestyle shifts for masses of people. The old occupations and professions will lose their value, which will be a very painful transition. In Great Britain, during the sixteenth and seventeenth centuries, the agrarian revolution generated many beggars and tramps, who were subject to harsh repression; the industrial revolution of the eighteenth and nineteenth centuries was accompanied by mass destitution among petty artisans and the suffering "reserve army of labor." Yet neither epoch saw a social catastrophe. Landless peasants either became agricultural workers for hire or were absorbed by the growth of industrial manufacturing. Artisans who went out of business filled the swiftly growing ranks of the industrial proletariat in the factories.

* * *

In the same way, *the coming technological revolution, even as it makes entire professions redundant, will also create new types of jobs.* New technologies will foster new needs, and meeting those needs will in turn summon new technologies. New jobs will appear to replace those that are liquidated by automatization and the growth of labor productivity. Moreover, inexorable growth in the relative weight of the "knowledge economy" (during the transitional stage), the rising need to acquire new knowledge, may absorb many workers.

* * *

As the technological basis of production changes and we transition to nooproduction, the very terms "profession" and "job" will drastically change their meaning, assuming they do not disappear outright. The word "profession," referring to a means of earning money in exchange for a certain type of skilled labor, will likely fade into the past. These functions will instead be performed by technetic beings, while humans liberate themselves from the narrow specialization of the present day. Thus, there will be no professions in the traditional sense: humans will focus on moving closer and closer to absolute knowledge, becoming ever more universally minded. New ways of accessing knowledge and information will be developed—varieties of neural networks, for example, and other human-machine systems.

* * *

Of course, human beings' noonomic versatility will not consist in every person knowing everything, but rather, in newly opened possibilities and newly acquired abilities for mastering practically any needed knowledge. *The main shift will consist of the creation of informational and communicative systems that allow any individual person to draw upon the entire ocean of knowledge accumulated by humanity, and to plumb this ocean's depths yet further.*

* * *

Needless to say, this will call for the improvement of the faculties of human beings themselves, who will need to master the ability to enter into any sphere of knowledge and orient themselves within it. This universal flexibility is fully obtainable so long as the education system and, yes, even human nature (remembering its limits?!) are restructured as needed. The main task of the education system will no longer be to "fill up" or "inflate" learners with knowledge and skills corresponding to a certain specialization. Learners must stop being passive customers, "hoarders," of ready-made knowledge, and instead learn to independently "extract" and use knowledge themselves. Of course, it will be impossible to acquire this ability without broad fundamental education to facilitate rapid self-orientation in any needed area of knowledge.

* * *

The transitional stage to such a vision of "universal self-learning humanity" will be the realization of the concepts of "education for all" and "lifelong learning," which are necessary for reaching the stage of NIS.2. Meanwhile, it is becoming critically important to research and develop new and progressively more advanced and versatile ways of accessing knowledge.

7.3. From Economic Rationality to Reasonable Formation of Needs

The fundamental shifts in the nature of people's activity and entire way of life that will occur during the transition to NIS.2 (followed by nooindustrial production) will be accompanied by substantial changes in the motivations for human behavior, that is, the structure and content of needs and desires.

Man, himself, will internally limit the uncontrolled inflation of such needs and desires. These transformations will be followed by reevaluation of the criteria for judging the rationality of production and of needs.

* * *

First and foremost, rationality implies correspondence to certain criteria. The framework of rationality is some set of criteria that we ourselves have established as a basis. How? Through knowledge: knowledge of certain things, comprehension of them, and the formulation of "border posts" that correspond to them, determining where one may reasonably go and where one may not.

* * *

This system of coordinates, or system of criteria, is dynamic. As we expand our knowledge, the space of rationality also expands, and so does our knowledge of what our basic criteria should be. As a result, our criteria and the boundaries of our rationality grow more expansive.

* * *

In the past, it was irrational and unreasonable for European women to leave half their legs bare, so they wore their skirts down to the floor. Why? Because of cold weather and the fact that since people spent most of their time either outside or in rooms with significant temperature fluctuations (with inconsistent heating, and sometimes not even that) there was always a risk of catching a cold. African women, by contrast, were able to walk around with their thighs barely covered by strips of cloth. On this basis, different norms, both moral and otherwise, arose in each place.

* * *

Later, this difference made its way into culture—behavior, clothing, culture at large—and coalesced as criteria for judging human action: a concept of what is "moral" and what is "immoral." The role of *culture as a phenomenon of society's development* is often seriously underestimated by adherents of strictly technocratic conceptions of social development. Contrary to this approach, however, it must be noted that the solutions to many problems generated by the turbulence of industrial and scientific progress can be found precisely in the cultural sphere, however strange this may seem at first. *Culture is where both society and individuals consolidate their view of what is rational—in production, in consumption, and in all of human behavior.*

* * *

It turns out, then, that rationality is always in flux; it develops along with society. At its heart lies humanity's capacity to take in ever more knowledge about how to satisfy needs, such as the need for fresh knowledge—including knowledge of "the good," whose boundaries can be shifted. Expanding one's rational framework is crucially important, as it allows us to understand not only how the world is structured, but why it seems to be "going crazy": after all, a shift in one's rational framework *does* mean "going crazy," in that it involves stepping outside one's prior rational boundaries. And for exactly this reason, all the things we have already spent many years studying come to seem totally unsuitable for analyzing or understanding the future, or for coming to terms with who we are.

* * *

With the opening of new horizons of knowledge (about everything!), many mental discoveries are later validated in practice. As a result, knowledge is verified, "enhanced"; our basic criteria are refined, and the expanse of our rationality is corrected, broadened, "moved."

* * *

In the past, it was unreasonable to wish for space flight. The desire to fly around the Earth was (according to the basic criteria of those past centuries) a fantastical, simulative need. For a present-day astronaut, this need is both realizable and not at all simulative, but it remains simulative (though realizable) for a typical space tourist: a wealthy potential tourist need only pay twenty million dollars to thirty million dollars for this experience. In the latter case, we might ask: Is this rational? Yet the day will come when everyone will fly in space, just as everyone now flies in airplanes, and this will be completely rational. In other words, over time, everything is gradually deformed and reformed, corrected, adapted; the space of rationality changes, along with its basic criteria.

* * *

This displacement and development of rationality's boundaries poses the question: What level of rationality should we aim for? Which criteria should we trust?

* * *

At any moment, we must trust the basic criteria within which we are situated. Knowledge, transferred into technology, makes possible the satisfaction of needs to greater and greater effect, giving us progressively more insight into how to comprehend the world, including comprehension of our own various needs:

simulative needs, non-simulative needs, the procedure for changing our needs, and so on. Accordingly, the basic criteria of a new space—a space of reason and rationality, moving us closer to non-simulative, reasonable, real needs—are formed.

* * *

The cognition of a more and more expansive rational space also implies moving away from certain values that were part of our basic criteria before. As soon as we "broaden our minds," we understand where the mistakes and faults in our criteria were. Once we correct these faults and move forward, we realize yet again that there are still certain inaccuracies, ambiguities, imprecisions, and that we must raise the level of veracity of our knowledge. Our basic criteria become more and more "true," bringing us closer to Absolute Knowledge, absolute belief. And sooner or later, drawing on a framework of belief in certain truths, society "slows down" and then "stops" the build-up of needs that are simulative at its current stage of development.

* * *

Knowledge demonstrates which needs are false and which are not. It reconstructs humanity, making it possible to believe that our basic criteria are credible and true and that we must "observe" these criteria, that is, live by them. We must accord with the level of understanding and self-comprehension that we have hitherto achieved. If we do not, it means that our basic criteria are still too narrow and must be expanded . . . In other words, simulative needs are negated by objective knowledge.

* * *

Knowledge, both of the external world and of ourselves, presumes the acceptance of limitations. In determining for myself who I am, I draw a boundary. If I define myself as a reasonable person, then by doing so I *impose a boundary on myself* that distinguishes me from an unreasonable person. Human beings are simultaneously characterized by movement towards previously unattained limits and by yearning to move beyond those limits. But this yearning is productive and creative, rather than destructive, only when it is regulated by the internal boundary that people establish for themselves.

* * *

To have a qualitatively new—nooindustrial—mechanism for satisfying needs, we will need a new reproductive link between production and consumption.

Humanity's needs, like the knowledge needed to fulfill them, will be shaped not by the process of direct productive activity (from which human beings will have departed) but by human creative self-development. We may refer to these needs as *noo-needs.*

> *Noo-needs* are needs determined by the criteria of human reason, and by cultural imperatives that are based on the satisfaction of vital needs at a rational level and on the escalating role of higher-order needs and desires.

* * *

Such needs and such knowledge, will produce a distinct type of "directive" or "technical task" for the autonomously functioning "human-free" sphere of direct material production. By transmitting this directive to the sphere of nooindustry, humanity will ultimately receive the necessary means to satisfy its needs without direct involvement in their organization or production. These tasks will be performed by a relatively autonomously functioning technosphere.

* * *

As the content of human activity and the way needs are satisfied change, so also will the criteria for determining the rationality of consumption change, along with the very structure of consumer needs. The criteria of economic rationality will be displaced by criteria of *reasonable* consumption, as defined by human culture. (See fig. 7.)

Humanity's orientation towards purely economic criteria of success is destined to fade away in the long term, and not only because human needs that draw on non-monetary values and motives, and which often cannot even be defined in terms of cost vs. effect, are on the ascent. Rather, economic rationality is becoming more and more dubious due to its own negative consequences. Such rationality deforms the structure of human needs, trying to fit them all into the Procrustean bed of monetary symbols of success; it takes only those achievements that stimulate the growth of market-derived measurements of value to be "rational."

* * *

In fact, should the incremental growth of value-based wealth always be considered beneficial, and should everything without monetary value be renounced as "irrational"? Practice teaches us that market prices fail to capture (or capture in highly distorted form) a great many important factors of human existence and development. Take, for example, such fundamental points as the dynamics of existence of our natural environment. Can the loss of biodiversity and the extinction of species be expressed in monetary terms? Is it possible to financially measure the level of human culture or the value of human communication?

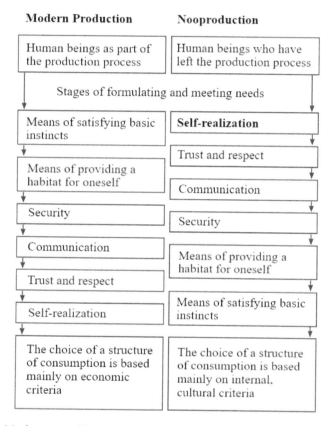

FIGURE 7. Mechanisms of formulating and meeting needs in modern production and in nooproduction

* * *

The marketization of everything is an ideology that does not stop at regarding human communication as merely a factor in the growth of capital. But does this mean that we should put an end to everything that does not generate surplus profit, and pronounce everything that does generate such profit as inherently valuable? If this is the point of view we adopt, then we can definitely say which path humanity will choose at the civilizational crossroads discussed earlier. Any need or desire, however false or perverse, will be placed at the forefront of human activity, so long as its satisfaction pays well. Where there is demand, supply will follow.

* * *

Humanity cannot be held by the hand and led away from this path. Rather, we must make a conscious internal choice, formulating new criteria of rationality based on the expansion of our field of knowledge and on the wealth of culture that we have created. This will be a step towards transitioning to the nooindustrial stage of development, in which our reason will acquire sovereignty over the elemental socio-economic processes that currently subjugate humanity.

Step Eight

From Economy to Noonomy

8.1. Replacing Economic Rationality—with What?

Human society is appropriately reacting to the growth in its technological ability to meet its non-material/cultural/spiritual needs by changing the trend of civilization's development. First and foremost, this means changing society's value systems, which requires changing the behavior of the human beings that uphold social values. Even the scholarly world is beginning to take note of these phenomena at last, though often only superficially and without an immersive understanding of their true meaning. Why did Richard Thaler win the Nobel Prize for Economics in October 2017? He won it for his acknowledgment of the fact that the economic behavior of most people (especially, of course, young people) is driven more and more by emotions rather than "rational" considerations.

* * *

Emotions are spiritual, immaterial, cultural value-bearing elements of the "average" person's overall structure of needs. This has always been the case. And humanity has always been driven to find ways of meeting its emotional needs, which dry economic rationality regards as far from trustworthy. The growth of Generation Z, who are much more advanced in this regard, is causing an increase in the importance of emotionally tinted needs within society's overall structure of needs.

* * *

This explains the evident fact that the decisions made by many "market actors" are less and less "rational"—within the framework of our current basic criteria of economic rationality, and from the perspective of dyed-in-the-wool apologists of the "animalistic" social reflection of humanity's biological essence. It will never occur to these market "generals" and "strategists" that the market is a vestige of the dying past, of the previous economic order, and that the observable (strengthening!) tendencies towards market "irrationality" are merely "sensors," detecting the rising transformation of humankind's preferences, as well as the decreasing importance of "rational marketplace behavior" and of the marketplace itself.

* * *

The Nobel Laureate's work, mentioned above, bears witness to the fact that some economists are finally catching on to the fact that people do not live according to the "indifference curves" found in economics textbooks. In fact, economists are lamenting that human beings, as it turns out, are not even capable of living this way! You see, there are "limitations" to human rationality . . . But perhaps this is a limited perspective on the problem? Man is not a "stupid ox," limited even in his market rationality. Quite the contrary: man's capacities are richer than that, and he is capable of making decisions guided by diverse criteria, some of which are non-economic. The goals of production, and people's core needs in general, are and have always been formed outside the marketplace, even in the very most capitalistic and market-driven societies.

* * *

A corollary of the development of noosociety, and of the shift towards nooproduction and noo-needs, is the transition away from economic rationality towards a new type of rationality. The new nature of rationality, corresponding to a new way of defining the targets of development, is of principal importance. In this way, economics will be displaced by **noonomy**, which is premised on moving away from the paradigm of growth as guided by economic "rationality" (defined as increasing volume price indicators) towards a developmental paradigm based on reaching concrete goals and on meeting humanity's real needs.

Our initial definition of *noonomy* is that it is a noneconomic means of organizing society's practical activity, oriented towards meeting humanity's concrete needs according to rational criteria defined by the development of knowledge and culture.

* * *

Both terminologically and substantively, the concept of noonomy echoes the idea of the noosphere.

* * *

The idea of the noosphere was first articulated by **Edouard Le Roy** (1870–1954), **Pierre Teilhard de Chardin** (1881–1955), and **Vladimir Ivanovich Vernadsky** (1863–1945). The development of this idea was spurred by Vernadsky's lectures on geochemistry, which he read at the Sorbonne from 1922–1923 to an audience that included Le Roy and Teilhard. Le Roy was the first to introduce the term "noosphere" into scientific discourse.[1] The noosphere was examined in greater detail in Teilhard's and Vernadsky's work at the end of the 1930s.

* * *

Pierre Teilhard de Chardin understood the noosphere as a qualitatively new concentration of consciousness, forming a special spiritual sphere, a "thinking layer" that covers the Earth. In his view, the concentration of thought at a planetary scale is closely tied to the convergence of individual human minds and spirits, which, as evolution continues, will lead to the emergence of "the Earth's spirit."[2]

* * *

V.I. Vernadsky took a more natural-scientific approach to the idea of the noosphere, arguing that humankind's rational activity was becoming the main transformative force with respect to both the biosphere and the Earth's geological shell (the biogeosphere).[3]

* * *

---

1 Edouard Le Roy, *L'exigence idéaliste et le fait de l'évolution* (Paris: Boivin & Cie, 1927).
2 See Y.Y. Novikov and B.G. Rezhabek, "The Contribution of E. Le Roy and P. Teilhard de Chardin to the Development of the Concept of the Noosphere," http://www.nffedorov.ru/w/images/3/36/Lerua.pdf.
3 V.I. Vernadsky, *Scientific Thought as a Planetary Phenomenon*, trans. B.A. Starostin (Moscow: Nongovernmental Ecological V.I. Vernadsky Foundation, 1997).

What we see in all these conceptions, however, is less a scientific theory than a range of interpretations of the uncontestable fact that the vital activity of human society and human beings, defined by their capacity for rational action, is becoming the decisive factor affecting not just humanity's own fate, but that of the Earth itself (or, at least, the Earth's surface).

* * *

The primacy of reason poses the inevitable problem of reason's own development. Which imperatives will govern *reason itself*? This, in turn, leads us to the question of how human society should be ordered to make sure that the power of reason is used reasonably, not merely as an effective instrument for satisfying our zoological instincts. The concept of the noosphere cannot answer this question.

* * *

The answer is provided by the notion of transitioning toward a noosocial order, toward *noosociety*. And the **noonomy** will be one of the basic elements of noosociety, serving as a kind of planet-wide "nomos" (meaning law, structure, order) that determines a non-economic mode of practical activity that meets human needs, using cultural imperatives to orient itself rather than economic rationality . . .

* * *

The term "noonomy" is derived from the Greek words "nous" (νους, reason) and "nomos" (νομός, order, law). Given that noonomy is defined in terms of practical management, it might seem logical to incorporate the Greek word "oikos" (οἶκος, home, household management) into this term. However, in the modern scientific tradition, terms formed from "oikos"—such as "economy"—are used to designate economic realities. Our intention is precisely to avoid identifying noonomy with any *economic* social structure.

Rather than mechanically combining the terms "noosphere" and "economy," we begin with an understanding of the Greek term "nous" as *reason*, resting upon the basic criterion of truth as a conscious, intrinsic value. Even in the eleventh century, Metropolitan Ilarion, in his "Sermon on Law and Grace," wrote: "He [the Lord] brought us unto the knowledge of the Truth."[4] In this sense, it would be deeply mistaken to equate the Greek word "nous" with its Latin analogue "ratio."

* * *

4 *Sermons and Rhetoric of Kievan Rus'*, trans. Simon Franklin (Cambridge, MA: Harvard Ukrainian Research Institute, 1991) 14.

To call something *rational* means to say that it corresponds to certain criteria (but are these criteria reasonable?) The economy is always rational, but do rationally acting economic subjects always act *reasonably*? And are they capable of moving beyond the criteria imposed on them by the current economic system?

* * *

Noonomy offers a different way to evaluate practical activity and needs, based not on rationality, but on reason, on "noo-," beginning with understanding the true consequences of practical decisions and the true value of the needs being met. Consequently, we are not dealing with economics, nor with individuals who rationally maximize their satisfaction, but instead with a different way of formulating and meeting needs, which we might call *noo-needs*.

* * *

Meanwhile, the second half of the term noonomy—"nomos"—is an ancient concept that was used in mid-twentieth-century philosophy to designate the basic organizational principle of any space, the absolute law of existence of all things.[5] In other words, noonomy is an orderly way of life, a mode of satisfying needs and desires in a society which cherishes the "light of reason" and has no place for production or relations of production; no place for property or property relations; and no place for economics—a society in which economics is impossible. *Noonomy is a non-economic system for meeting noo-needs.* For this reason, it is wrong to speak of the "economy of the noosphere," just as it would be to refer to non-economic economies, non-predatory predators, and so on.

* * *

In a market economy, rationality is understood only in terms of maximizing monetary gain. Of course, neoclassical economic theory rejects the charge of reducing economics to money, claiming that people are prone to try to maximize their receipt of all kinds of goods; in reality, though, these goods only come to economists' attention once they obtain a monetary value.

* * *

It is only relatively recently that the neoclassicists, under pressure from research in behavioral economics, have somewhat softened their position. They now acknowledge that human beings are not a pre-programmed calculator of profit and loss, that people can be moved by other motives, and that economic

---

5   See Carl Schmitt, *The Nomos of the Earth in the International Law of Jus Publicum Europaeum*, trans. G.L. Ulmen (Candor, NY: Telos Press, 2006).

decisions can be influenced by noneconomic factors. Yet they integrate all this into their thinking by explaining it away as humanity's "limited rationality." "Genuine" rationality is still conceived in terms of measuring profit and loss, but human beings, alas, are imperfect, and their capacity for rational behavior is limited by various circumstantial factors.

* * *

Indeed, for capitalist market economies, this attitude is largely (though not entirely!) justified, since it meets the basic criteria of capitalist rationality. However, changing the social conditions of production leads to changing the rational criteria of human behavior (see Fig. 8). The transition to nooproduction and noonomy leads to the criteria of reason displacing the criteria of monetary gain, as it becomes rational to orient oneself towards satisfying concrete, reasonable needs and desires. The needs for knowledge, for trust, for social recognition, and for self-realization become dominant over the rapacious need for material goods, and increasing the volume of goods one consumes ceases to be a basic goal of human activity—precisely insofar as this need has already been sated within reasonable limits.

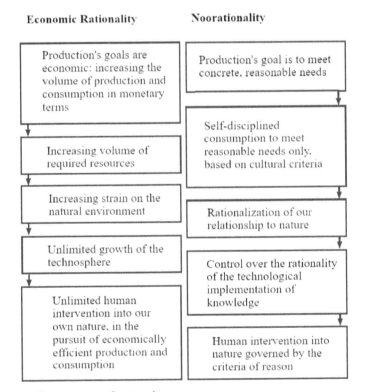

FIGURE 8. Different types of rationality

* * *

These goals also inform the building of new mechanisms to regulate nooproduction, which will be oriented not towards "noo-GDP" or profit, but towards different indicators that reflect what we want to achieve. To meet this task, incoming flows of data, management information, material, and more will be formed. Such flows and control actions should be planned out and programmed in the quantity, time, place, and frequency required to reach a desired result.

* * *

Thus, noonomy prioritizes not the private pursuit of profit or other gain, to be obtained through the chaotic play of market forces, but the rational effort to *meet concrete needs that have been evaluated as reasonable*. Accordingly, the level at which these reasonable needs are to be met amounts to a concrete goal of production.

* * *

This assumes a specific program of action that rises above the chaos of markets, giving production a more systematic, planned, orderly character. Such an approach cannot exclude the element of randomness, nor the freedom of individual choice unlimited by prescriptions imposed from above. It follows that the program of production we develop must be significantly flexible and adaptive to changing circumstances and chance disturbances.

* * *

Next, this production program must be corrected if anything goes awry, since the number of factors that the plan must account for is much greater than what we can analyze at our existing level of knowledge.

* * *

Let us take one (artificial) example. I host a TV program and invite my colleague to the studio. There are two glasses on a table in front of us, and we have planned to drink water from these glasses. Then we are told that we should come back to the show in one year and get 100 percent more than what we've already achieved—that is, two more glasses. Will we need these two extra glasses? No. But once they have been made, poured, and set up in the studio, our GDP will have doubled!

* * *

This is an artificial example of the absurdity that may, in its destructive force, bring all civilization to the point of catastrophe if we continue down this path. In fact, this approach, which is often supported today by both business structures and various state programs, does nothing but propagate simulative things. Previously, this same approach was supported by both the Soviet system, in its own way, and the non-Soviet system, in a different way; both systems fostered the simulative orientation of development, "growth-oriented" economic development, without either "cutting out" the illusory and false components of society's structure of needs or conceptualizing the goals of economic planning to this end.

* * *

For this reason, we may formulate the principle of the economy (still an economy) of the future, which is already dawning, as follows: *we do not need linear economic growth, we need economic development.* In this regard, growth is in fact a fiction.

* * *

Based on such a viewpoint, the indicators that are used today in attempts to quantitatively measure our happiness (that is, to "measure harmony with algebra")[6] should be consigned to the dustbin of history. We need other criteria—a new and different set of basic criteria—within which society's development might be qualitatively evaluated.

* * *

To this end, obviously, we must implement a more rational (according to our new basic criteria) mechanism of planning. This could take many forms; we will not insist on specific methods.

* * *

The main thing, the basic thing, is to satisfy people's real needs. It is crucial to comprehend and evaluate emergent non-simulative needs. If the market gives rise to an enormous quantity of fictive needs, what is to be done with them? Banning them would be stupid, not to mention impossible. But if we do not ban them, the economy may become a fiction, evaporating into hot air, and a decent future might be squandered.

---

6   The author quotes a famous line from Alexander Pushkin's short 1830 play *Mozart and Salieri.*
    —*Translator's note*

* * *

What can we do under these conditions? Evidently, we need a system of thoughtful behaviors and stimuli, but not just economic ones: if we are transitioning to this new paradigm, then the economy as such, in its current, observable form, will no longer work. An "entirely new normal" will be emerging. Here, we must strive not to orient ourselves using traditional economic indicators, but instead to consider which individual and social needs we should satisfy: which of them, exactly, will propel us most effectively (with the fewest possible costs and conflicts, at the highest speed, etc.) towards NIS.2 and beyond?

* * *

Only once we are capable of meeting these needs can we say that "happiness has grown." That *happiness* has grown—not GDP, but happiness. This task is much less trivial than merely planning economic growth, as our economic authorities currently do (and for fairness' sake, we should acknowledge that this is true outside of Russia as well). However, if humanity grasps the importance of this task, and if the task becomes a conscious, real need, then in our view, the problem can be resolved even at our current scientific and technological level.

* * *

Hardly anyone will argue that happiness does not lie in inflating GDP, or profit, or greater accumulation of money. It is both funny and sad whenever someone seriously, not as a joke, proposes that "happiness does not lie in money, but in how much of it you have," or when people's disinclination to chase after money is attributed to their "limited rationality." After all, human rationality is not defined as the choice of exclusively economic "gains."

* * *

Humanity is more intelligent and rational than is supposed by these ideologues of quantitatively measured growth. People do not just need to acquire new phones, or another glass of water, or anything else; they need glass of a certain quality, water with a particular taste, and "a certain quality" of life itself. Perhaps the truth is that two glasses are unnecessary: one is enough, so long as it is *good*, beautiful, comfortable, and contains clean water.

* * *

Precisely this small nuance—"is it good?"—is the most important thing. The behavior that now gets called "irrational" amounts to picking just one of the two glasses, even though both are being imposed on us. Or, better still, we are told to smash the old glasses and throw them away, in favor of three new ones, "for the price of two." We could take two glasses—this would be more, this would be growth. Yet we choose just one glass, because we like it more. We like it not for any transcendental reason, nor because the glass is some illusory thing, but because there are internal parameters we use to make a judgment: size, for example (perhaps we choose a cup because our finger fits through the handle), or because we rationally perceive and evaluate some aspect of the object as beautiful.

* * *

In fact, this is a different kind of rationality, a different kind of knowledge, a different kind of reason. In fact, our reason, our rationality, is far more wide-reaching and richly endowed than the economic boundaries within which today's economic paradigm tries to enclose us.

* * *

Relatedly, we may observe that even in today's world—a world of developed market economies, shot through with narrow economic rationality—a significant proportion of goods are distributed for free. This fact points to an important tendency: the farther we go, and the more society's transition to the next age of industry (regulating the value of the products and services we produce) accelerates, the more goods will be freely available.

* * *

For this reason, it is time for us to leave the paradigm of economic growth behind and start making use of "growth-oriented" parameters only for subordinate purposes. It is time to include social consciousness in the formulation of our economic model and of new conceptions of civilizational, economic, and social development. This is because economics and society are inseparably linked. Long ago, during Soviet times, people talked about "socio-economic development"; I would prefer to speak of "economico-social development." But what is development? It is the gradual rejection of everything that creates today's simulative economy. This is, first and foremost, a transition in our "economic minds."

* * *

This is why measuring society's development exclusively according to narrow markers such as GDP or other quantitative macroeconomic indicators, which are not equivalent to society itself, is inadequate and even unscientific. We must find other parameters and establish fitting goals for economic planning. Where we should look for them is no secret: we should look for them by meeting people's genuine needs. That is, we should make judgments not according to purely physical methods of measurement, but through qualitative measures: surveys, focus groups, and other such indirect research methods. Nowadays, new technologies—"big data"—provide instruments for this type of analysis. It is time to move from the arithmetic of trivial addition to "calculus," even though the latter is more complicated.

* * *

At one point, intellectuals belonging to the Club of Rome articulated the idea that we must limit economic growth in order to prevent ecological catastrophe. Intelligent people! Of course, what they meant was slightly different: they were proposing to reduce our impact on the biogeosphere by lowering consumption. We can agree that limiting consumption may to some extent (though not for certain!) reduce our environmental footprint, but our position differs in principle: we must reduce consumption of *simulacra*, while more fully satisfying genuine needs.

* * *

Economics, which focuses on quantitative measurements and on the creation of ever-newer energies, products, and things without considering their genuine necessity, is leading us down a dead-end path. Humanity needs a different economy. Or rather, a different "-nomy" altogether, no longer an economy, corresponding to the needs of the world that is our home.

* * *

Recall the alter-globalization movement's famous slogan: "People, not Profit." "Need, not Greed."[7] In the twenty-first century, this has basically become the main slogan of worldwide social forums.[8]

* * *

---

7 Sasha Simic, "Need, not greed," *The Guardian*, January 25, 2007, https://www.theguardian. com/commentisfree/2007/jan/25/post997.

8 See, e.g., World Social Forum 2016, Global Justice Now, https://www.globaljustice.org. uk/event/world-social-forum-2016/; Praful Bidwai, "A Great Movement Is Born: Global Justice Movement Finds Fertile Ground at the Asia Social Forum," Focus on the Global South, January 28, 2003, https://focusweb.org/a-great-movement-is-born-global-justice- movement-finds-fertile-ground-at-the-asia-social-forum/.

From a noo-perspective, this is an entirely positive slogan. Not because its proclaimers are do-gooders or "revolutionaries": the right tone for this slogan would not be *for* revolution, but against it, advocating gradual, evolutionary, orderly, and reasonable development. The theoretical platform outlined in this book perceives the proper meaning of this slogan with crystal clarity: money is nothing but an intermediary, so it must inevitably recede, while humanity is of principal importance. Therefore—"people, not money."

8.2. A Non-Economic Mode for Regulating Practical Activity

The shaping of a non-economic modes of practical activity will occur as we move from today's economic order towards NIS.2, and through NIS.2 to noonomy.

* * *

Here we can distinguish two stages of the process of historical movement. The *first stage* involves the development of *"technologies of trust,"* making it possible for economic relationships between people (within which satisfaction of needs takes place) to involve collaboration without an intermediary (monetary or otherwise). This is the basis for "compressing" the economic forms of human activity, the economic institutions that mediate the link between production and consumption.

* * *

The *second stage* involves the *disappearance of labor effort itself as the mediating link between human beings and the satisfaction of their needs* (see Fig. 9.) The Old Testament's maxim— "By the sweat of your brow you will eat your food"—will recede into the past. Thus, both the nature of human activity and the means of satisfying human needs will both fundamentally change, becoming noneconomic. In a certain sense, humanity will return to "paradise," approaching the Absolute. Or, to put it another way, humanity will enter Marx's "realm of freedom."

* * *

During the first stage, then, we remain within the sphere of economics and production relations, but with technologies already emergent that can minimize today's world of economic relations. "Technologies of trust," plus accelerating overall technological progress, will constrict the overblown sphere of intermediaries, provision of transaction operations, etc.

Stage I — Development of technologies of trust

Compression of relations of intermediation

Compression of economic forms and institutions mediating the satisfaction of needs

Stage II — Disappearance of labor as an activity that mediates the satisfaction of needs

Disappearance of economic relations (money, capital, property, etc.)

FIGURE 9. Two stages of movement towards noonomy

* * *

At the second stage, the need for people to serve as mediators for the satisfaction of other people's needs disappears entirely. To put it crudely, neither the baker nor the shop salesman will meet our need for bread; rather, this need will be met by the bakery, all by itself. The same applies to masses of other professions. Interaction between people will continue only in the process of creative activity, the process of discovering new knowledge and "transmitting" it to the technosphere, implementing it in new technologies.

* * *

But even before the formation of nooproduction, *creative activity that implements knowledge in new technologies will effectively change our collective mode of acquisition.*

* * *

The substantial difference between the acquisition of knowledge and the acquisition of material products consists in the fact that knowledge, once it has been obtained, can never be "recovered" or "extracted" away from its new owners. With material objects, everything is more straightforward: they can be taken and then given back. But knowledge cannot be "irrevocably" returned.

* * *

But the expanded application of knowledge influences the acquisition of material products as well as intellectual ones. With the development of new knowledge and new technologies, the easier, cheaper, and simpler obtainment of material goods brings with it less and less need for intellectual private property—in fact, less and less need for property itself as an institution: not for knowledge, but specifically for property.

* * *

What future is there for the information-based, "knowledge-based" aspect of production? However hard we might try to limit the use of the results of scientific research with artificial rules, sooner or later these results will "seep through," manifesting themselves in social production and the social order to form new social conditions. In other words, sooner or later, our struggle with scientific research will come to an end. But right now, we are at the first stage of the formation of a long and great transition.

* * *

On one hand, this is the beginning of profound comprehension of the value of knowledge as the future's most important resource. On the other hand, today's prevailing social relations are founded on private means of appropriation of the results of social production and competition over the resources needed for production. Consequently, they give rise to those means of "protecting" intellectual property that temporally "prolong" currently existing social relations with respect to knowledge, taking the relations that have emerged in the "material" sphere and disseminating them within the sphere of knowledge. Of course, this stage will be overcome with the development of NIS.2.

* * *

Even during NIS.2, tendencies will emerge that lead to the transformation—and even the withering away—of the economic forms of human activity. This is especially visible in activities that involve the acquisition of new knowledge. But what will replace these economic forms? Won't the sphere of production remain, though without humanity's direct participation? Won't the sphere of humanity's creative, "knowledge-producing" and "culture-producing" activity, entirely uninfluenced by social relations, also remain?

* * *

Here numerous general questions arise. How will people organize their influence on human-free production? How will the goals and orientation of production be decided? What aspects of production must be managed and regulated? After all, though the sphere of production will exist outside of human relations, it will not exist independently of people: no less than before, the reproduction of human life will depend on it!

* * *

At this point, human development confronts a dilemma. Either society will be unable to direct the capacities unleashed by the technological revolution toward its own improvement and will be captivated by false goals and values, exacerbating the negative tendencies of contemporary civilization to the point that humanity loses its very essence—which would indicate that we will not enter noosociety or transition to noocivilization. Or, alternately, humanity will successfully enact a noo-approach to redesigning its current civilizational attitudes.

* * *

Nooproduction, though it will be carried out independently from human beings and from society, will remain subordinate to society in its goals and tasks. Precisely the sphere of setting goals, formulating goals and tasks, and regulating the accessible means within the technosphere for realizing these tasks—all this will remain within the purview of human social relations. Autonomous techno-beings, functioning within the sphere of nooproduction and capable of self-development, will nevertheless depend on human society, which will define the limits of their self-development and block any developmental directions that are not of social use, pointing the functioning and development of nooproduction in directions that are necessary for the development of humanity itself (see Fig. 10.)

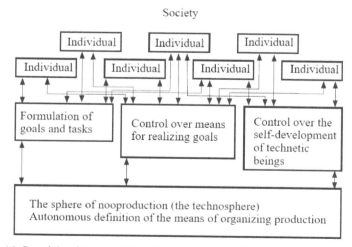

FIGURE 10. People's relations within the process of regulating nooproduction

* * *

This will involve incomparably more profound shifts than merely, for example, taking account of ecological limits when making economic decisions. It will mark the beginnings of qualitative transformations in the content of production, in the needs, values, and motivations of human behavior, and naturally also in socioeconomic relations and institutions. Let us repeat that the basis for this will be created by qualitatively new technologies that convert the semi-utopian modalities of the twentieth century into the practically realizable tasks of the present.

* * *

We are not remotely inclined to idealize either the theory of the noosphere or the phenomenon that is now emerging before our very eyes—that is, noosociety, along with its distinct forms of production and its new practical-material relations, which are no longer economic in the strict sense of the word. For Vernadsky, the noosphere is not intentionally created; it is an inevitable product of human society at a particular stage of its self-development. In itself, the noosphere *does not guarantee* the "kingdom of truth." Therefore, the presence of the noosphere already poses the question as to *precisely which* imperatives of reason will prevail within it.

* * *

Hence arise the questions and challenges to which we must respond, such as the question of the *social form of nooproduction.* Which imperatives will govern both the production of the material and spiritual conditions of human life, and the social relations that regulate this production? What determines the choice of such imperatives? To a significant extent, the condition of the noosphere as a whole will depend on the answers to these questions.

* * *

We have already given a first, approximate response, emphasizing the necessity for cultural imperatives to come to the fore in shaping needs and regulating practical activity oriented towards satisfying needs. We designated the social form of such regulation using the term *noonomy*, which we may now define in greater detail.

> Our expanded definition of *noonomy* is that it is a noneconomic social form of human practical activity, aimed at satisfying noo-needs (especially those involving the development of human personality) at a particular developmental stage: that of *nooproduction*, which is the form of production that characterizes

humanity's exit from direct labor activity ("human-free production"), as well as management of the technosphere as a sphere that is external to humanity but dedicated to realizing the potential of human cognition.

* * *

The means of governing social relations will become different, and governance itself will take on a new meaning, more consensual and different from what it is today. Even though we may continue to call this system of governance a "state," it will be a qualitatively different type of state than it was before. The main difference between the state in economic society and the state of the future is that today's state, like the ones before it, primarily regulates economic relations. In the future, economic relations will disappear along with the economy itself, but other types of relations will remain, which will require their own kind of regulator.

* * *

I believe that there ought to be institutions of social regulation, and that the anarchist ideal of complete self-regulation is untenable. We need a system of relations that allows us somehow to learn about others' interests and juxtapose them with our own. We need basic criteria for making decisions. The basic criteria I have in mind are cultural: culture's criteria constantly evolve as development proceeds. This is to say that we need appraisals of some kind about how quickly to develop, what path to take, and so forth and so on.

These appraisals will amount to a certain means for reaching consensus, a consensus form of social government. Since society is constituted by different interests, people's individual interests exist alongside shared social interests. As these interests develop further, the need for consensus-based regulation will also develop—but now according to cultural rather than economic criteria, based on the power of human reason: noocriteria.

It stands to reason that social bonds will remain, since these are what embed human beings within a community. But will such bonds retain the character of social relations, that is, relations between people as elements of a social structure, as representatives of social classes and social or professional groups, etc.? We may suppose that these types of social relations, too, will wither away: noonomy will dispense with the foundation for the separation of people into classes and professions (along with the withering away of professions themselves), and generally with divisions according to social status.

* * *

Moreover, provision of the material conditions of existence will itself cease to be directly performed by human hands. Marx's prognosis about the displacement of human beings from the process of material production will come true. Humanity will influence production not through manual effort, but through the power of knowledge.

8.3. Human Society and the Autonomous Technosphere

In order to comprehend the social arrangement of production in noosociety—that is, noonomy itself—let us briefly review the structure of production.

* * *

The production process is the process of manufacturing products, that is, the transformation of natural substances for the purposes of meeting human needs. The most essential elements of the production process are human *labor*, original *materials* (*including raw materials*), *technologies*, and the *organization* of production.

* * *

> All production rests upon the following four basic points, which can be most easily articulated in the form of four questions:
> *What are our ingredients?* (Original and raw materials.)
> *What are our instruments?* (Tools and technology.)
> *How are we working?* (The content and character of labor.)
> *How will we organize our work?* (Organization and regulation of production.)

* * *

These elements change as their knowledge-density increases, changing the basis of the production process, which further influences (or is reflected in) society and its socioeconomic institutions. In turn, social relations and institutions influence the elements of the production process, either facilitating or hindering their development.

* * *

The interaction of these components in the process of manufacturing the products we need is of utmost importance. The current developmental trend towards the increased role of knowledge in all components of human activity within the production process (in tools and technologies, in the labor process itself, in the organization and management of the production process) has already been noted repeatedly in the previous pages, along with the corresponding trend for the relative share of material resources consumed in production to decrease. So, what will become of the social form of production as we transition to the noosocietal stage?

* * *

The form of productive organization based on knowledge-intensive technologies will change significantly even in NIS.2, since it will directly depend on the character of the technologies it uses, shedding production's former "factory" form (the management of groups of people that attend to systems of machines). Since people will be increasingly displaced from the process of direct production, a special role will be played by forms of human social interaction—*within the bounds* of the production process but oriented towards regulating production.

* * *

The development of nooproduction will trigger an even more significant shift. In the new world, the production process' form of organization will cease to play any special role, since this form will automatically arise from the self-development of the nootechnological space without any direct participation by human beings.

* * *

However, the transition to "human-free production" presents us with a highly complex task, though not from the vantage point of technological problems.

* * *

As humanity develops, we have gradually become more and more distant from nature, placing an intermediary between nature and ourselves: namely, labor activity, which relies on technologies whose progress is based on ever-increasing knowledge about nature. The general developmental trend of human society was to progressively reduce our direct dependence on nature through the development of the technosphere.

* * *

Humanity tried to progressively reduce the risks that nature subjected us to in our primitive state, always striving to move as far as possible away from a natural world that posed us every imaginable threat. Humanity warmed itself, clothed itself, protected itself with walls, housed itself, armed itself against everything that threatened it, created stockpiles of resources to survive natural cataclysms and changes in its external environmental circumstances, and so on. By distancing/distinguishing ourselves from nature, but never forgetting to gather its fruits, we reduced the entropy of our existence, the risk of being frozen, eaten, crushed by a falling tree . . .

* * *

Humanity used more and more refined means to process and "reshape" natural resources to fulfill our own needs. At the same time, we spatially separated ourselves from our natural surroundings, erecting houses and entire cities with many millions of inhabitants and artificial life-supporting systems, practically to the point of complete separation from natural conditions.

> Saudi Arabia is among the top global consumers of fresh water per capita (926 cubic meters per person per year) despite the fact that only 9% of its water comes from "natural," renewable sources. Water pipes extend through the desert, from city to city, for hundreds of kilometers. Meanwhile, Persian Gulf states like Bahrain, Qatar, and Kuwait primarily use desalinated sweater (accounting for 69%–79% of their overall consumption).[9]

* * *

In fact, the whole history of the development of our civilization is a history of our separation from nature. Here we must acknowledge the dualism of the situation. Human beings, having realized ourselves as personalities and thereby spiritually separated ourselves from the surrounding world, have nonetheless remained part of nature—indeed, precisely this "separation" leads us to a further separation between "spiritual" man and "natural" or "biological" man! This gulf can take the dramatic form of reckless incursions into nature, both external nature (environmental transformation) and our own (interference in humanity's biological essence).

* * *

---

9 PwC Strategy&, *Achieving a Sustainable Water Sector in the GCC: Managing Supply and Demand, Building Institutions*, Strategy&, 2014.

We are now approaching a threshold at which humanity is not just separated from nature, but ready to separate itself also from "second nature," from the artificial world created by our own hands.

* * *

The sixth technological paradigm, based on NBICS-convergent technologies, is already creating sufficient preconditions for transitioning to nooproduction, which signifies *the final separation of the technosphere from human society*. Separation not in the sense that humanity will lose all links with the technosphere or cease to reap its rewards, but in the sense that humanity's direct participation will no longer be necessary for the technosphere to function. The link between noosociety and nooproduction will remain, but now as a kind of "bottleneck": a channel of interaction rather than the embeddedness of one within the other.

* * *

At the same time, this separation is quite dialectical. Not only does humanity nearly separate itself from the surrounding world, including the technosphere, through knowledge and the comprehension of natural forces (and social forces); to a certain extent, and precisely due to this separation from the technosphere, humanity returns to nature. Henceforth, rather than "subjugating nature" in the traditional sense, humanity turns natural processes towards its own ends—but not at the cost of blind encroachment upon nature without consideration for the harm we do to it.

* * *

No, it stands to reason that "the struggle with nature," in the sense of making nature serve human ends, will remain. But this struggle will become more "technological," more thoughtful, more reliant on cognition of nature, thereby making humanity's interaction with nature (including our own) ever more harmonious rather than mutually destructive. What was once a genuine fight will become a kind of partnership. This conversion of natural processes into technological processes, which relies on knowledge and the new values and cultural imperatives produced by knowledge, can harmonize the links in the chain running from the natural environment (biosphere), to production, to humanity.

* * *

Now, having separated itself from its natural basis, humanity need not act like a predator and force its way into nature. This implies, among other things, a more careful and tactful attitude towards human essence itself, with limits placed on thoughtless incursions into the human organism to reconstruct it based on fleeting impulses of the present moment.

* * *

It is precisely on this basis that Vernadsky's theories about the noosphere (which we understand as the sphere of activity of "noo-humanity") will be realized; without the harmonization and sublation of the conflict described above, these ideas are unrealizable.

8.4. The Role of Humanity in Noonomy

Thus, the new technological paradigm's achievements will be able to genuinely lead humanity, for the first time ever, beyond the boundaries of direct material production. The production that will be established on this foundation can already be called nooproduction, in the sense that human reason and knowledge will serve as both its defining resource and its main regulator.

* * *

Karl Marx already foresaw this future in the second half of the nineteenth century, presciently discovering it in the tendency for the role of human knowledge to grow in the development of industrial production.[10] But only now, and for the first time, can we more or less exactly locate the concrete technological basis that will truly allow humankind to exit material production, while remaining (in Marx's terms) its "watchman and regulator."

* * *

Such a fundamental shift in society's technological base will bring with it a no less fundamental shift in social relations. If humanity exits the direct process of production, then the relations created by humanity's productive activity will disappear as well. Productive relations will gradually disappear, and production

---

10 Marx noted the tendency for "the transformation of the production process from the simple labor process into a scientific process, which subjugates the forces of nature and compels them to work in the service of human needs" (*Grundrisse*, trans. Nicolaus, 700); that is, production's transformation from labor into "experimental science, materially creative and objectifying science" (*Grundrisse*, trans. Nicolaus, 712).

will lose the form of economic activity. Economics and the economy will become a thing of the past.

* * *

Of course, nature abhors a vacuum. Though humanity will leave production, the latter will remain, as before, the key material condition for human life, and people will find some way to establish social relations among themselves in order to regulate the production process. But since these relations will not, in themselves, be directly included in the production process, *the results will no longer be economic, but noonomic: not social relations within the framework of production that is carried out directly by humans, but relations concerning nooproduction, which develops without direct human participation but is regulated and directed by human reason.*

* * *

Factually speaking, what this means is that practical activity that meets human needs will no longer primarily be defined using economic criteria, since needs will themselves take on noneconomic form. Moreover, the economy, as the sphere of economic relations between people concerning production and the exchange of products, will generally shrink to the point of disappearing entirely. Not, let us add, because production's raw material or energy costs will cease to be important, but because humanity will no longer be directly incorporated into productive activity, meaning that the corresponding relations between people will no longer emerge.

* * *

Humanity will depart from direct production, making the technetic entities spawned by the technosphere do all production's necessary work . . . The economy will become superfluous. Practical and productive processes will become a thing-in-itself that no longer interests us. People will move beyond the boundaries of this process.

* * *

In this case, it is most important to note the following principle: unlike all previous stages, the essence of the noo-stage of civilizational development is that individuals no longer relate to one another within the process of material production; rather, it is two differently constructed civilizational spheres that enter into relations with one another—production (nooproduction, confined to the technosphere) and human society. (See Fig. 11.)

Modern Civilization

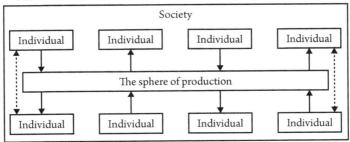

The Age of Noonomy

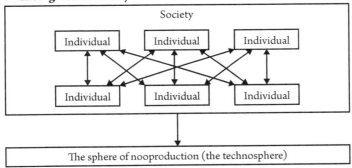

FIGURE 11. The transformation of civilization's design during the transition to noonomy

* * *

Thus, the structure of human civilization—civilization at large, not just society—will change in principle. Society here stands out as one isolated part of human civilization, since, for the first time in history, the technosphere will distinguish itself (in a certain sense) from human society. Of course, society and the technosphere did not directly coincide before either, but previously, people constructed their social ties with one another according to their direct incorporation into the functioning of the technosphere. Now, by contrast, people will relate with human-free material production as its "watchman and regulator."

* * *

In this light, the question emerges as to what human activity will look like. I propose that humans will "administer activity," which presupposes that humanity constantly and consciously thinks about what it is doing, an approach which must become a way of life rather than a mere skill. Just as technology transfer will become an inextricable and ceaseless element of life rather than a mere episodic "introduction" of new technologies, humanity's way of life must come to incorporate our administration of ourselves and of society through each person's interaction with everyone else.

* * *

But does this mean the disappearance of production relations between people? Yes, as people exit the sphere of direct production, the relations of direct production will disappear. But what about relations between people defined by their influence on the sphere of "human-free production," determining the developmental path of such human-free production? Presumably, those relations will exist, won't they?

* * *

Yes, but these will no longer be the specifically economic relations that we already know, and their role will resemble the roles of other types of activity that are entirely unrelated to production. Meanwhile, the specifically economic forms of social life will gradually wither away.

* * *

One of the key economic forms—property relations—will progressively lose their significance (due to the increasing accessibility of goods and the decreased value of owning them). First, this will occur within the sphere of people's practical and productive activity; later, property relations will completely depart from all people's social relations. With the disappearance of property, the economy as such will also disappear, since relationships of appropriation and alienation constitute the basis of economics.

* * *

Why will the significance of property plummet? Because the value of what property gives us will decrease, since anyone will be able to obtain any non-simulative thing that they need—and the farther we progress towards NIS.2, the more easily, simply, quickly, etc. this will be possible. Anyone, without effort, will be able to catch a fish from the pond. Not only without effort, but also without the advantages obtained through labor, namely the appropriation of its results as property.

* * *

Our technological level is gradually approaching a line which, when crossed, will allow all our rational needs to be satisfied more and more cheaply. Why cheaply? Because, in effect, everything humanity has done, all that humanity has procured for the realization of its needs, was obtained through a) natural resources, which have no monetary cost (!), and b) knowledge, obtained and "expended" so that original natural resources (not to mention "non-natural" materials, which are really just "natural" materials that have been "polished" using knowledge!) can, with the "addition" of knowledge, be transformed into something else.

* * *

Thus, everything that humanity has done, all its technological conversions— all of this is knowledge. Everything new is obtained through added knowledge; there is nothing else to it. Besides, materiality is free, while the share of knowledge in produced products grows ever larger (due to the new stages of processing accumulated with every new product or "iteration" of an existing product). Products become more and more expensive, and they would grow "infinitely" expensive if knowledge were owned exclusively. But in fact, the nature of knowledge is that while it may be exclusive at the moment of its "appearance" in the world, it immediately begins to spread itself far and wide and, consequently, to grow cheaper. The more that the exclusivity of knowledge goes down, the cheaper knowledge becomes.

* * *

In today's terminology, this is the "distribution of costs" for "information products," based on an extreme case of increasing returns to scale due to the extremely low cost of duplicating and disseminating information.[11] If "prices" begin to fall as we progress towards NIS.2, then it also becomes cheaper—ever more so—to "mine" products. In other words, what we have here resembles the economic concept of exponentially declining value. Hence, the meaning of property will also decline, as it will no longer "cost" anything!

* * *

Today, property consists of "materialized labor." A reserve of trade, for the satisfaction of future needs. Humankind began to appropriate goods as soon as it developed an "economic" conception of life: we need to have reserves, we need

---

11 W. Brian Arthur, "Increasing Returns and the New World of Business," *Harvard Business Review* 74, no. 4 (Jul-Aug 1996): 100–109.

this and that, we need it for our family, we need to keep it from our neighbors; we need to carve out our own little piece, protect it from the rest of the world, and announce that we won't give it to anyone else. That is all "property" is: an element of the most predatory manifestation of human nature.

* * *

But in the future, goods will "materialize" without labor. Reserves will be unnecessary. What will then be the point of "property?" When property is unnecessary (precisely property as a reserve or stockpile, because, of course, noonomy means that anyone can get any non-simulative thing they need) then what could appropriation possibly mean? Why would there be any need to appropriate? The word "mine" will even vanish from our lexicon, in the sense that nothing will be "mine" any longer—it will be merely a part of the world that fulfills reasonable human needs without today's concept of "labor" as traditionally understood. So, it will be impossible to refer to my own mountain of goods; there will merely be *a* mountain of goods, pure and simple.

* * *

Not only will technological progress change people's social relations and their mode of interaction with the technosphere, but it will also impact human nature itself. The pause in human evolution caused by the cessation of the mechanism of natural (biological) selection, linked to the fact that the adaptive significance of human beings' natural traits has been evened out by their capacity to make use of technology, will probably soon come to an end. Human evolution that is not biological but properly "technological," will begin—not the evolution of the technosphere, nor merely the evolution of humanity's habitat, but rather the technological evolution of human essence itself.

* * *

Won't a unique form of "technological selection" then emerge to replace natural selection? This question is already emerging as we observe the broad incursion of artificial intelligence (AI) systems into human life. Though, at present, AI systems are mostly used in business applications, they are bound to penetrate other spheres as well—science, education, medicine, the social sphere . . . The interaction of AI systems with humanity in these spheres will lead, as its next step, to a new stage of human evolution.

* * *

One can imagine the transformation of human nature and humanity's conversion into a neo-biotechnological species. Such beings will be a product of the synthesis of, on one hand, humanity's directed evolution as a species due to biological decisions which do not infringe upon humanity's essence as biological beings, and, secondly, humanity's technetic evolution due to its "reconstruction" by non-biological technologies.

* * *

The history of the development of human civilization is, at the same time, the history of the development of its technosphere: the "dead" world, whose technological "species" nonetheless undergo evolution in the same manner as the living world. This "dead world" also "lives" its own form of "life." This world includes the growth of a variety of technetic "species" and the formation of "technocenoses"—"habitats," "survival zones," "distribution zones," processes of "adaptation" and "acclimatization"[12]—as opposed to the narrowing diversity of biological species, the simplification of biotas, and the degradation of the biosphere as a whole.

* * *

Although, before the creation of AI, this "life" was animated by technology's creator—humanity—with the creation of AI it has become autonomous. AI technologies feature extremely high knowledge "content," including knowledge about how to "extract" and use new knowledge. This is why the question of "bookmarking" regulators ("commandments"?) within the tools used by AI, preventing unsanctioned implementation of knowledge that might harm humankind, is of principal importance. The development and constant improvement of such regulators should become one of humanity's most important tasks in the twenty-first century.

* * *

The evolution of the technosphere, and the techno-trends shaped by it, lead us to the question of the limits of civilization's development—insofar as technological evolution is beginning to set the parameters of human evolution, both in terms of evolution's biological, "material" basis and in terms of its social qualities. The phenomenon of artificial "selection" is emerging (less in the sense

---

12 The concept of "technocenosis" was coined by Boris Ivanovich Kudrin. See B.I. Kudrin, "Studies of Technical Systems as Communities of Technocenosis Pieces," in *System Studies. Methodological Issues. Annals 1980* (Moscow: Nauka, 1981): 236–254.

of a struggle for existence than as the search for, and selection of, cultivated human qualities), which will become a factor in humanity's "technological evolution."

* * *

Technologies for "editing" the genome are becoming accessible, providing a means for "selecting" people even before their birth, along with technologies for embedding auxiliary technetic elements into our biological essence, which continues the process of "selection" after birth. Technologies for "rearing" people will also change, altering our genome with the goal of "touching up" our neurochemical mechanisms of regulating behavior, to the point of various methods for influencing the consciousness of already fully formed human beings. Finally, even the means of "producing people" may change: the first steps are already being taken towards technologies for artificially growing highly organized living organisms outside of mothers' bodies . . .

* * *

All these emergent technological possibilities, "doors," and "windows," through which we can glimpse movement towards a new civilization (still human?), must be evaluated both in terms of the risk of civilizational crisis and from the vantage point of realizing the prospects of noospheric society. If we adopt a "noo-evaluation" of these trends, we must clearly perceive the contradictions that we may confront on the road to the future.

* * *

Soon we will need to bypass crises stemming from the misuse of technological interference into human essence. We cannot even precisely foretell what consequences such careless interference might bring about, or what varieties of no-longer-human species it might generate, or what relations will take shape between these species.

* * *

We can predict the appearance of a new type of social inequality—based no longer on property, but on unequal access to knowledge and unequal ability to master it. And now we should ask: how can the problems of such new inequality be resolved? Won't some people fight to turn their intellectual abilities into a new basis for social privilege, and won't an "anti-intellectual" wave emerge as a reaction to this?

* * *

The society of the future will be able to cope with these conflicts insofar as knowledge becomes the primary resource of production, and insofar as broader access to the mastery of knowledge, and to creative activity based on knowledge, becomes the necessary condition of society's development. In this light, even leftover inequality of property will be not a basis for social privilege, but instead the object of efforts to develop the creative abilities of every person as fully as possible.

* * *

Of course, we must not forget that *both knowledge and technology can be used to humanity's detriment and turned towards our self-destruction.* But knowledge and technology are also capable of ensuring humanity's *transcendence of all objective limits; the guaranteed resolution of problems that appear insoluble; and the ability to overcome barriers that seem insurmountable.*

Conclusion

The Path Towards
Nootransformation

———————

* * *

Everything said above leads us to the conclusion that movement towards noonomy is an objective tendency determined by transformations in the material conditions of production. However, human civilization faces many problems and contradictions that will have to be overcome along the way, and we will have to pass through the crossroads of civilization by making the right choice.

* * *

This choice determines whether our society will embark along the road to transforming its system of social relations into noosociety and its economic subsystem into noonomy. If we do not surmount the looming crisis of our civilization, the problems that are now menacing us threaten to grow into catastrophes.

* * *

Society should approach this civilizational crossroads, not just having armed itself with knowledge of its developmental prospects, nor merely having worked out genuine criteria for making such a collective and public decision, but also drawing on necessary material and technological prerequisites. By this, we refer to the creation of the fundamentals of the second generation of new industrial society.

* * *

If we turn from general reflections on the transition to NIS.2 to specific consideration of the prospects that await Russia on this path, we must begin with a clear understanding of Russia's place in the system of global relations. It makes sense here to transition to a broader frame of analysis, using the approach of world-systems theory and the achievements of geopolitical economy.[1]

* * *

More and more in the modern world, we observe the potential long-term subjugation of countries that do not manage to possess advanced technologies within about thirty to forty to fifty years of their appearance, due to the lack of institutions and instruments allowing them to establish technological parity or even—at least in certain sectors—become technological leaders.

* * *

In that case, if the current trend of separation into "center" and "periphery" continues, countries will be sorted into two groups. One group, it follows, will be the "productive" countries; that is, the new capitalists, figuratively speaking, who possess the capital of the future, namely knowledge and technology. The other group will be the "service" countries, working for their "crust of bread," for better or worse.

* * *

This pattern is combined with the current trend of social Darwinism and the "bio-paradigm" of social development. We have already discussed above where this trend might lead. Overcoming it is an urgent task, a collective societal requirement for global civilization. Relatedly, there must also be a counterweight to this trend, at least in the form of an alternative for the thinking part of humankind.

* * *

---

1 On the world-systems approach, see: Immanuel Wallerstein, *World-Systems Analysis: Theory and Methodology* (Beverly Hills: Sage, 1982); Samir Amin, *Unequal Development: An Essay on the Social Formations of Peripheral Capitalism*, trans. Brian Pierce (New York: Monthly Review Press, 1976). See also, Radhika Desai, *Geopolitical Economy: After US Hegemony, Globalization and Empire* (London: Pluto Press, 2013).

The Soviet Union might have become such an alternative to a certain extent if it had preserved itself and gained access to modern technology. But due to gigantic, unresolved internal and external problems, the USSR disintegrated without changing the course of world history in this way. Yet the continuation of our current form of "development," without an alternative, promises the collapse of civilization.[2]

* * *

Thus, when discussing Russia, it would be desirable to view the idea of its industrial renaissance not just as a way to avoid falling behind in economic competition with the leaders of the world-system (though this, too, is quite important during the stage of transition to NIS.2!), but as the creation of an alternative model that could bring about the world-system's transformation in line with a noo-version of development.

* * *

Where do things stand right now? How prepared is Russia to play this role? Our current situation is such that we fully risk falling into the second category of countries, even though we have the prerequisites we need to avoid falling behind. Look at our Soviet and subsequent Russian background; the mentality of our people; Russia's abilities in many areas; and the fact that we could obtain substantial income from our current natural resources and make well-directed investments. We have a decent scientific, industrial, and technological background, and we are capable of ending up in the first group of countries so long as we ensure our advancement towards NIS.2.

* * *

If we do not follow this path, then there will be nothing to discuss, and we entirely risk ending up in the lagging group of "service countries." We must clearly consider, understand, and recognize this fact.

* * *

Our country can claim sophisticated technological achievements, even on an incontestably global scale, and tangible improvements have begun in the real economy's high-tech industrial sectors. In recent years, after the government announced a "crusade" for a digital future, we have turned to economic

---

2   Samir Amin, *Russia and the Long Transition from Capitalism to Socialism* (New York: Monthly Review, 2016).

digitalization. But overall, the main thing is missing: there is no clear movement towards reindustrialization or the rapid modernization of our economy.

* * *

For many years, we have discussed the necessity of *reindustrializing on a qualitatively new technological foundation*—the foundation of our new economic model, that is, the necessity of restoring industry's preeminence in a qualitatively new way. Why do we talk about this so much? Because a new wave of technological transformations is coming, a new industrial and technological revolution, and its leaders will be those who manage to ride this "ninth wave."

* * *

Relatedly, I must emphasize the very important fact that those in the West who are now studying the problem of development are also carrying out economic reindustrialization in their own countries—even despite the substantially higher standard of Western industry!

* * *

Right now, *a "window of opportunities" has opened for Russia*, connected to the fact that the world capitalist economy, in its current model, is slowing down qualitative, revolutionary shifts in its own technological foundation. The advancement of sixth-paradigm technologies looks impressive, but these technologies are still quite far off from transforming the face of modern production. The world is evidently moving toward a technological revolution, but while this movement is happening quickly, its tempo is inadequate and its rhythm is ragged.

* * *

A small amount of time still remains: the modern economy's core trend of chasing after economic results is largely oriented towards not technological progress, but the return of huge profits at the expense of inflating financial bubbles. Up until this point, the world has been developing according to the following model: "we'll eat more, we'll drink more, and we'll cheat everyone on the financial market."

* * *

Of course, betting on a technological breakthrough does not prevent us from actively exporting gas during the present stage. Only fools would decline to make use of their advantages. But the income obtained from these exports must be deliberately invested in our economy's scientific-educational-industrial component.

* * *

As we shape the development strategy of our national economy, we must consider some basic tendencies of NIS.2, particularly the relative decline in industry's "raw-materials-density." We must clearly understand that in the coming decades, global demand for traditional raw materials, resources, and energy—hitherto the basis of the Russian economy—will tend to fall, not rise. This is inevitable given the dramatic rise in the role of industrial knowledge and technologies, and the rates at which these are obtained, mastered, implemented in the real sector, developed, etc. Natural resources will mean much less. That said, we not only "can" but must use our natural resources!

* * *

It is the changing proportion of materials-density versus "knowledge-density" in productive output that allows us to expect that the country we pass on to future generations will not be stripped of natural resources or strewn with spreading, cancerous waste dumps. But to this end, Russia must acquire advanced technologies.

* * *

As I have stated, we have no choice. Or rather, we do have one, but it is quite harsh: either we wrest our way out of our current position and into the ranks of the world's technological leaders within the next twenty years, or we will become part of the "periphery." In the latter case, we will "spoon-feed" the fruits of production, created through barbaric exploitation of our natural and human resources, to more developed countries.

* * *

To this end, *the Russian economy's systemic qualities need to change,* and quite substantially. There should be a transition to long-term strategic economic management, medium-turn indicative plans and programs based on scientific prognoses, and active industrial policy. The state should sponsor businesses' investment in R&D and technological refitting. It should also guarantee a stable and supportive tax situation and accessible and comfortable lending programs for the real sector, especially its high-tech fields. Admittedly, this system might foster a moderate degree of social differentiation: still, citizens' income should mostly depend above all on their real contribution to the economy.

* * *

Bearing in mind our goals, we can see that our current economic model contains practically no "escalators" or "elevators" that might boost our upward progress—just a few small movements. We are seriously late. For good reason, the economic community, and even the President, are now discussing the need to create a new model for turning our economy around. The old model has exhausted itself under present conditions and cannot go on any longer without leading to major problems, or even national catastrophe. Nonetheless, no particularly active moves have been made to build the "escalators" we need. We have no apparent plan to meet this challenge.

* * *

Planning should be appreciated as a phenomenon of a generally higher order than chaos, given its ability to decrease entropy and bring the dynamics of systemic development into order. When compared to the absence of a plan, or to the market, in this regard, planning is a step forward, capable of attaining a higher level of stability for the socio-economic system. The development of civilization moves along a path of increasing planned elements of economic development. And this is natural, though of course it does not proceed without missteps, sidesteps, and historical pitfalls—which, incidentally, are also natural, caused by society's lack of comprehension of the importance of planning.

* * *

It is becoming clearer and clearer that *we need to return to planning*. This is demonstrated by the experience of China, which did not reject planning as an institution and as an instrument for managing development. By applying the tools of planning, China's economy and society have moved much more quickly towards a new type of industrial society than, for example, we have done by repudiating planning and accepting chaotic principles of economic decision-making, creating conditions in which the strongest and most self-interested actors win out over social needs. Arguably, the intellectual society of the future, noosociety, is inconceivable without instituting planning as one of the basic, primary instruments of public administration and of society's very being.

* * *

Finally, without this instrument. we will not manage to seriously reach the next step—the problem of digitalizing the economy, that is, establishing modern information technology as an economic foundation, rather than turning digitalization into a fleeting fad or an empty slogan. After all, the most important infrastructural element of the modern economy is its informational component.

* * *

Today—already today, not tomorrow—our level of knowledge and our technological solutions are advanced enough to help us solve many problems. For this reason, we can talk about the possibility of more effective planning: selective planning, indicative planning, or really any type of planning that would make it possible to combine markets with the plan, as is constantly discussed from Russia to China to Scandinavia.

But how can we get to this point?

* * *

By combining modern communicative and information systems with the abilities of cognitive technology, artificial intelligence, self-learning systems, human-machine systems, etc., we can create the possibility of "digitalizing" both planned and market-based approaches to optimizing economic decisions, and also enable the integration of these two approaches with one another.

* * *

Of course, in establishing this new technological base for economic calculations, we will need to improve the economy's institutional structure, as this structure makes it possible to effectively orient the economy toward reindustrialization and create the material foundation for a technological breakthrough to the future.

* * *

There are two challenges that we urgently need to meet.

The first is the conscious adoption of a new model for our economy that presumes the priority of industrial development, along with all the decisions—economic, institutional, and so forth—that stem from this premise.

And the second is the consolidation of society, and of our national elites, to realize this model on the basis of shared responsibility to resolve the pressing issues faced by our country.

* * *

Russia is *already late* jumping onto the NIS.2 train. To avoid ending up in the group of "catch-up" countries, we must bet on the most promising directions of development, "beyond the horizon." Of course, these will be mere daydreams unless we exert extraordinary effort to gain access to NIS.2-level technologies—if we want to get ahead of the curve. *Building the noonomy must start now,* even if at a very narrow level; we must, in practice, work painstakingly to manifest *everything that is bound to become the future of human development and economic activity.*